THE TOXICOLOGY OF METHANOL

THE TOXICOLOGY OF METHANOL

Edited by

John J. Clary

Bio Risk, Midland, MI, USA

WILEY

A JOHN WILEY & SONS, INC., PUBLICATION

Cover Design: John Wiley & Sons, Inc.
Cover Illustration: © Urbanglimpses/iStockphoto

Published by John Wiley & Sons, Inc., Hoboken, New Jersey.

Published simultaneously in Canada.

Library of Congress Cataloging-in-Publication Data:

The toxicology of methanol/edited by John J. Clary.
p. cm.
Includes bibliographical references and index.
ISBN 978-0-470-31759-4 (cloth)
1. Methanol--Toxicology. 2. Methanol--Environmental aspects. I. Clary, John J., 1937-
RA1242.W8T69 2012
615.9'02--dc23

2012023644

10 9 8 7 6 5 4 3 2 1

CONTENTS

Preface xiii

Contributors xv

1 Methanol Production and Markets: Past, Present, and Future 1

Gregory A. Dolan

References 9

2 Methanol: Fate and Transport in the Environment 11

Rula A. Deeb, Todd L. Anderson, Michael C. Kavanaugh, and Lauren A. Kell

2.1 Introduction 11

2.1.1 Release Scenarios 11

2.1.2 Fate in the Environment 14

2.2 Partitioning of Methanol in the Environment 16

2.2.1 Methanol Partitioning Among Environmental Media 16

2.2.2 Air/Water Partitioning 16

2.2.3 Soil/Water Partitioning 18

2.2.4 Methanol Dissolution 19

2.2.5 Commingling/Cosolvency Effects 21

2.3 Fate and Transport of Methanol in the Environment 22

2.3.1 Soil and/or Groundwater Release 23

2.3.1.1 Sources of Methanol in Soil and Groundwater 23

2.3.1.2 Losses of Methanol from Soil and Groundwater 23

2.3.1.3 Methanol/BTEX Commingled Plumes 32

2.3.2 Surface Water Release 33
2.3.2.1 Sources of Methanol in Surface Water 33
2.3.2.2 Losses of Methanol in Surface Water 33
2.4 Methanol Additives 37
2.4.1 Luminosity 38
2.4.2 Taste 38
2.5 Conclusions 40
References 40

3 Human Toxicity **47**

John J. Clary

3.1 Introduction 47
3.2 Exposure 48
3.2.1 Dietary 48
3.2.2 Environmental 49
3.3 Metabolism in Humans 50
3.3.1 Normal 50
3.3.2 High Exposure 51
3.3.3 Over Exposure 51
3.3.3.1 Symptoms 51
3.3.3.2 Blood and Urine Methanol 52
3.3.3.3 Urinary Formic Acid 53
3.3.3.4 Breath—Methanol Levels 53
3.4 History of Human Toxicity 54
3.4.1 Occupational 54
3.4.2 Ingestion 54
3.4.3 Dermal 59
3.5 Controlled Human Studies 60
3.6 In Utero Exposure 62
3.7 Repeat Inhalation Exposure 63
3.8 Management of Methanol Poisoning 64
3.9 Conclusions 66
References 67

4 General Animal and Aquatic Toxicity **73**

John J. Clary

4.1 Introduction 73
4.2 Acute Toxicity 74
4.2.1 Oral 74
4.2.2 Dermal 78
4.2.3 Inhalation 79
4.2.4 Intraperitoneal 82
4.2.5 Subcutaneous 82
4.2.6 Intravenous 82
4.2.7 Other Acute Studies 82
4.3 Irritation 86
4.3.1 Dermal 86
4.3.2 Eye 87
4.4 Sensitization 87
4.5 Repeat Exposure—Inhalation 87
4.5.1 Non-Human Primates 87
4.5.2 Rats 91
4.5.3 Mice 93
4.5.4 Dogs 94
4.6 Repeat Exposure—Oral 94
4.6.1 Rats 94
4.6.2 Non-Human Primates 95
4.6.3 Mice 95
4.7 Repeat Exposure—Dermal 96
4.7.1 Mice 96
4.8 Aquatic Toxicity 96
4.9 Conclusion 99
References 100

5 Developmental and Reproductive Toxicology of Methanol **107**

John M. Rogers, Jeffrey S. Gift, and Stanley Barone, Jr.

5.1 Introduction 107
5.2 Reproductive Toxicity 108

5.3 Developmental Toxicity 110
5.3.1 Rats 111
5.3.2 Mice 115
5.3.3 Non-Human Primates 120
5.3.4 Summary of Developmental Toxicity Findings in Experimental Animals Exposed to Methanol by Inhalation 124
5.3.5 Pathogenesis of Methanol-Induced Birth Defects 127
5.3.5.1 Whole Animal Studies 127
5.3.5.2 *In Vitro* Studies 128
5.3.6 Folate Deficiency—A Susceptibility Factor for Methanol Developmental Toxicity? 129
5.3.7 Role of Methanol and Metabolites in the Developmental Toxicity of Methanol 133
5.4 Conclusions 136
Disclaimer 139
References 139

6 Exploring Differences Between PBPK Models of Methanol Disposition in Mice and Humans: Important Lessons Learned 145

Thomas B. Starr

6.1 Background 145
6.2 Are Humans More or Less Sensitive than Mice to the Toxic Effects of Methanol? 148
6.3 Are the Two Models' Predictions of Human Blood Methanol Concentrations at Steady State Consistent with Each Other? 153
6.4 Are the Values of Key Human Metabolism Parameters Consistent with Those in the Published Scientific Literature? 155
6.5 Shouldn't the Possibility of Systematic Bias be Considered Carefully During the Model Fitting and Parameter Estimation Process? 160

6.6 Is "Visual Optimization" an Adequate Technique for Estimating PBPK Model Parameters? 161
6.7 When Human Data are Available, Shouldn't they be Utilized in Making an Objective Comparison of Model-Specific Predictions? 163
6.8 Summary of Lessons Learned 164
References 165

7 Oxidative Stress and Species Differences in the Metabolism, Developmental Toxicity, and Carcinogenic Potential of Methanol and Ethanol 169

Peter G. Wells, Gordon P. McCallum, Lutfiya Miller, Michelle Siu, and J. Nicole Sweeting

7.1 Introduction 169
7.1.1 Preamble 169
7.1.1.1 The Regulatory Problem 169
7.1.1.2 Fundamental Question 169
7.1.1.3 Research Objectives 170
7.1.1.4 Approach 170
7.1.2 Methanol Developmental Toxicity 172
7.1.3 Carcinogenic Potential 176
7.1.4 Oxidative Stress and Other Potential Mechanisms of Toxicity 177
7.1.5 Factors Affecting the Human Relevance of Animal Models 178
7.1.5.1 Species Differences in Metabolism 178
7.1.5.2 Dose of Methanol and Route of Exposure 179
7.2 Species Differences in Methanol Metabolism 179
7.2.1 Enzymes and Pathways 179
7.2.1.1 Alcohol Dehydrogenase (ADH1) 179
7.2.1.2 Catalase 182
7.2.1.3 Cytochrome P450 (CYP) 2E1 184

7.2.1.4 Formaldehyde Dehydrogenase (ADH3) 186
7.2.1.5 Folate-dependent dehydrogenase 187
7.2.2 Pharmacokinetics of Methanol and Formic Acid 188
7.3 Species and Strain Differences in Methanol Toxicity 191
7.3.1 Acute Metabolic Acidosis, Ocular Toxicity, and Death 191
7.3.2 Teratogenesis 194
7.3.3 Neurodevelopmental Effects 204
7.3.4 Carcinogenic Potential 207
7.4 Oxidative Stress 213
7.4.1 Oxidative Stress Mechanisms 213
7.4.1.1 Embryonic Drug Exposure and Reactive Oxygen Species (ROS) Formation 213
7.4.1.2 Signal Transduction 214
7.4.1.3 Macromolecular Damage 217
7.4.2 Oxidative Stress from Methanol Exposure 223
7.4.2.1 Evidence for MeOH-Initiated ROS Formation 223
7.4.2.2 Mechanism of MeOH-Initiated ROS Formation 227
7.4.3 Teratogenicity of Methanol and Comparisons to Ethanol 228
7.4.3.1 Genetic Modulation of Catalase 228
7.4.3.2 Free Radical Spin Trapping Agent 232
7.4.4 Carcinogenic Potential 233
7.4.4.1 Oxidatively Damaged DNA 233
7.4.4.2 Hydroxynonenal-Histidine Protein Adducts 234
7.5 Conclusions 237
Acknowledgment 238
References 238

8 Methanol and Cancer **255**

John J. Clary

8.1 Introduction 255
8.2 Rodent Bioassay 256
8.2.1 Oral 256
8.2.1.1 Rats 256
8.2.1.2 Mice 263
8.2.2 Inhalation 266
8.2.2.1 Rats 266
8.2.2.2 Mice 268
8.2.3 Dermal 268
8.2.3.1 Mice 268
8.3 Possible Mechanisms 270
8.3.1 Genotoxicity 270
8.3.1.1 *In Vitro* 270
8.3.1.2 *In Vivo* 270
8.3.2 Oxidative Damage 272
8.4 Human Cancer Data 276
8.5 Conclusion 276
References 277

Index **283**

PREFACE

Methanol is a large volume chemical. It is widely used as a chemical intermediate and in many consumer products. It has been in commercial use for over 100 years. It is the simplest alcohol. It is used in paints, plywood subfloors certain types of carpets, windshield washer fluid, antifreeze, in other automotive products, in gasoline blend, as a stand-alone automotive fuel, and in the production of biodiesel. Methanol is water-soluble and it will quickly biodegrade in the environment. It is used as a building block for formaldehyde, acetic acid, and other chemical derivatives. It is naturally found in our diet (fruits and artificial sweeteners). Uses have changed and new uses are being developed.

This book presents the toxicity of this widely used chemical both in humans and animals as well as examining the difference between species. Information comes from experimental studies in animals and human experience. The majority of the human data is from acute exposures. Methanol can cause blindness and death following large oral acute exposures. The animal data, which goes to making up the majority of the data on the toxicity of methanol and the mechanism of action, have been reviewed as it relates to the potential toxicity in humans.

CONTRIBUTORS

Todd L. Anderson, ARCADIS U.S., Inc., Emeryville, CA, USA

Stanley Barone, Jr., National Center for Environmental Assessment, United States Environmental Protection Agency, Research Triangle Park, NC, USA

John J. Clary, Bio Risk, Midland, MI, USA

Rula A. Deeb, ARCADIS U.S., Inc., Emeryville, CA, USA

Gregory A. Dolan, Methanol Institute, Arlington, VA, USA

Jeffrey S. Gift, Hazardous Pollutant Assessment Group, National Center for Environmental Assessment, Office of Research and Development, United States Environmental Protection Agency, Research Triangle Park, NC, USA

Michael C. Kavanaugh, ARCADIS U.S., Inc., Emeryville, CA, USA

Lauren A. Kell, ARCADIS U.S., Inc., Emeryville, CA, USA

Gordon P. McCallum, Division of Biomolecular Sciences, Faculty of Pharmacy, University of Toronto, Toronto, Ontario, Canada

Lutfiya Miller, Department of Pharmacology and Toxicology, Faculty of Medicine,University of Toronto, Toronto, Ontario, Canada

John M. Rogers, Toxicity Assessment Division, National Health and Environmental Effects Research Laboratory, Office of Research and Development, United States Environmental Protection Agency, Research Triangle Park, NC, USA

Michelle Siu, Division of Biomolecular Sciences, Faculty of Pharmacy, University of Toronto, Toronto, Ontario, Canada

Thomas B. Starr, TBS Associates, Raleigh, NC, USA

J. Nicole Sweeting, Division of Biomolecular Sciences, Faculty of Pharmacy, University of Toronto, Toronto, Ontario, Canada

Peter G. Wells, Division of Biomolecular Sciences, Faculty of Pharmacy; Department of Pharmacology and Toxicology, Faculty of Medicine, University of Toronto, Toronto, Ontario, Canada

1 Methanol Production and Markets: Past, Present, and Future

GREGORY A. DOLAN

Methanol Institute, Arlington, VA, USA

Methanol has a long proud history dating back to ancient times when Egyptians formed it as a byproduct of charcoal fabrication from wood (Crocco, 2002), which was then used to preserve mummies. Not much changed in the intervening centuries to improve the process. In 1923, world methanol production stood at just 30,000 tons (1 ton of methanol contains 333 gallons), distilled from 3 million tons of wood feedstock. In the same year, Matthias Pier of BASF produced the first railcar load of synthetic methanol from a converted ammonia loop. In post-World War II Germany, methanol was produced from petroleum liquids and coal for fuel use. In the 1960s and 1970s, companies such as ICI in the United Kingdom and Lurgi in Germany began developing specialized catalysts for methanol synthesis from natural gas in low-pressure processes. Over the next two decades, the methanol industry would grow from a "captive" market with plants located next to their downstream derivative (i.e., formaldehyde or acetic acid typically) to a global "merchant" market, with methanol widely exported around the world.

In 2011, world methanol demand topped 45 million tons (CMAI, 2011) and with 65% of this consumption being traded from one continent to another, methanol is clearly one of the world's most widely distributed chemical commodities. Owing to the steady growth of

The Toxicology of Methanol, First Edition. Edited by John J. Clary.

methanol demand, we have seen a significant rebalancing of methanol production. Referred to in the industry as a "rationalization," the plants in regions with rapidly increasing natural gas feedstock costs have been closed, as new "mega" methanol plants are built in countries where natural gas is more plentiful and less expensive. These "mega" methanol plants have capacities of 5000 tons per day (600 million gallons per year), with a single plant representing close to 5% of global production. Production capacity in North America and Western Europe fell from 13.3 million tons in 1999 to just 900,000 tons in 2010. During this same time period, production capacity jumped from 13.1 million tons to over 24.5 million tons in South America (led by Trinidad and Tobago) and the Middle East. The real wild card in the global methanol industry is China, which saw the production capacity soaring from just 1.2 million tons in 1999 to 40 million tons by 2011. By 2007, China had become the world's largest methanol producer and consumer, with the breakneck pace of new methanol plant construction building further momentum for growth.

Today, we are seeing the pendulum beginning to swing back again. We are now seeing a re-emergence of North American methanol capacity, driven by the increasing availability of shale gas and its impact on pushing natural gas prices below $4 per MMBtu. Formerly, mothballed methanol plants in Canada and Texas have been restarted, with one of these facilities recouping its restart costs in just 7 months. One major producer is looking to ship one or two methanol plants from South America, which have had challenges accessing natural gas, to the U.S. gulf coast, which now boasts the lowest costs for available natural gas in the world market.

Today, most of the world's methanol production comes from the steam reformation of natural gas, characterized by the two-step equation:

$$CH_4 + 0.5\,O_2 \rightarrow CO + 2\,H_2,$$

$$CO + 2\,H_2 \rightarrow CH_3OH.$$

The methanol production process involves four basis steps (Figure 1.1): (1) feed gas purification to remove natural gas components such as sulfur that can poison catalysts; (2) steam reforming to saturate the

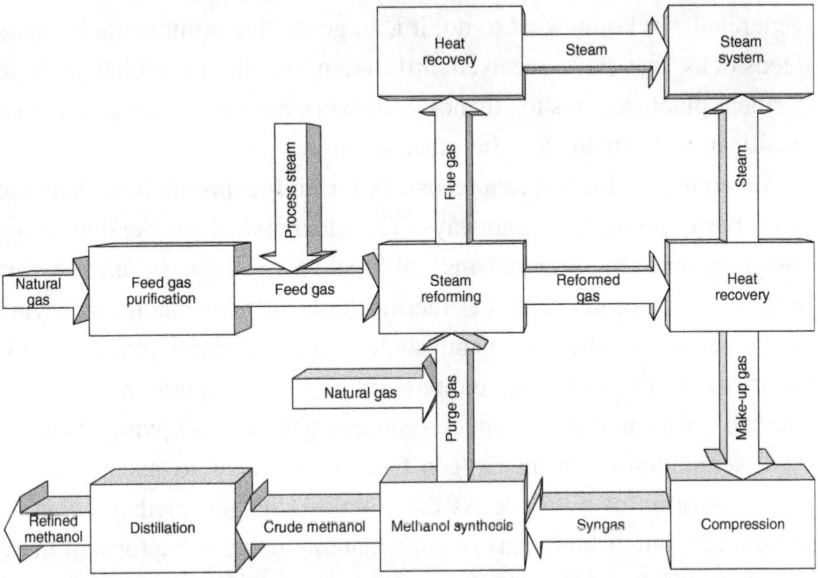

FIGURE 1.1 Conventional methanol production.

hydrocarbons producing a synthesis gas of carbon dioxide and hydrogen; (3) methanol synthesis by passing the synthesis gas over a catalyst bed at high temperatures and pressures to produce crude liquid methanol; and (4) distillation typically accomplished in a two-step process to remove water and some ethanol created in the process. The finished methanol must meet rigorous purity standards generally in the order of 99.85% (ASTM D-1152/97).

The production of methanol from natural gas, coal, or biomass shares a number of basic processing steps (Zuberbühler, 2005). The feedstock must be gasified by heating in the presence of little or no oxygen to produce a synthesis gas made up of carbon monoxide, hydrogen, carbon dioxide, and water (along with various other gases). This "syngas" is then catalytically processed into liquid methanol while much of the "equipment" for gasification involves mature technologies using recognized feedstocks. While majority of methanol is produced through the steam reformation of natural gas, China has focused on converting its vast coal resources to methanol via gasification. For "biomethanol," the immature part of the equation is the first step, the gasification of biomass (a feedstock with different characteristics). Once the syngas is generated, we know what to do; it is to get to that point using biomass feedstocks that have received little attention. In the global push to ferment plant starches to ethanol, little work has been done on biomass gasification to methanol, the other alcohol.

When using biomass as a feedstock for biofuel production, there are four basic production pathways: (1) biochemical conversion using enzymes and microorganisms to breakdown biomass into sugars used for fuel production; (2) thermochemical conversion employing heat energy and chemical catalysts to convert biomass into fuels; (3) gasification to dissociate biomass in a high-temperature, oxygen-starved environment to produce synthesis gas; and (4) pyrolysis using high temperatures in an oxygen-free environment to encourage the decomposition of biomass. As the simplest alcohol, methanol can be produced from virtually any organic material using some form of these processes. However, the most common methods employed to produce methanol from biomass involve the gasification of "dry biomass" (forest thinnings, waste wood, pulp mill byproducts, municipal solid waste) and the fermentation of "wet" biomass (animal manure, waste-water, industrial wastewater, algae, seaweed) typically through anaerobic digestion (Specht and Bandi).

Biomass gasification for methanol production is especially attractive as high carbon conversion rates and fuel yields mean that the biomass resource can be completely utilized. By comparison, conventional production processes for the biochemical conversion of plant starch and oil plants use only a small fraction of the biomass feedstock. For example, it is understood that production of ethanol from corn yields 7.2 dry tons/ha/year, or 76 GJ/ha/year, whereas the production of methanol from wood yields 15 dry tons/ha/year or the equivalent of 177 GJ/ha/year (Williams et al., 1995). In other words, through gasification, 1 ton of woody biomass can produce 165 gallons of methanol while the hoped for yields for cellulosic ethanol is targeted to around 60–70 gallons of fuel per ton of biomass. As the Swedish Minister for Enterprise and Energy Deputy Prime Minister Maud Olofsson put it, "We need to move away from first generation in ethanol manufacturing and further to second and third generation, which is all about cellulose material and gasification, and this implies therefore room for methanol and synthetic diesel" (Olofsson, 2012).

Further, the production of methanol from biomass gasification may turn out to be an evolutionary stepping-stone to a "fourth" generation technology. In his seminal text "Beyond Oil and Gas: The Methanol Economy" (Olah, 2006), the Nobel Prize Laureate Dr George A. Olah of the University of Southern California argues that we may soon recycle atmospheric carbon dioxide using catalytic and electrochemical methods to produce liquid methanol. As Dr. Olah states, "It should be emphasized that the 'Methanol Economy' is not producing energy. In the form of liquid methanol, it only stores energy more conveniently and safely compared to extremely difficult to handle and highly volatile hydrogen gas, the basis of the 'hydrogen economy'. Besides being a most convenient energy storage material and a suitable transportation fuel, methanol can also be catalytically converted to ethylene and/or propylene, the building blocks of synthetic hydrocarbons and their products, which are currently obtained from our diminishing oil and gas resources."

This is an important point, as the petrochemical industry has grown hand in hand with the petroleum industry for good and bad. Methanol is a basic building block for hundreds of chemical commodities such as formaldehyde and acetic acid used in products ranging from building materials and plastics to paints, adhesives, and solvents. We even color methanol blue for the windshield wash fluid in your car today.

As a chemical building block, methanol is a key component of hundreds of products that touch our daily lives. The largest global market for methanol is as a feedstock for the production of formaldehyde. Engineered woods used in building our homes and furniture are bonded with resins produced from formaldehyde. In our cars, urethanes and plastics used in essential components also contain formaldehyde. Methanol is also used in the production of acetic acid, which then is used for making polyethylene teraphthalate (PET) plastic used in beverage packaging. Acetic acid is the basic component of terapthalic acid (PTA), which is used in making polyester fiber for our clothing and carpets. Vinyl acetage monomer (VAM) is also produced for acetic acid and is used in the manufacture of water-based paints and adhesives. On a global basis, the fuel additive methyl tertiary butyl ether (MTBE) is still used to increase octane performance and reduce emissions in vehicles. MTBE is produced from methanol and butanes and continues to play an important role as a fuel oxygenate in Asia and the Middle East.

As can be seen in Figure 1.2, methanol is both an important chemical commodity and an energy fuel. The "Other" category includes several applications for consumer products that are widely familiar, including windshield wash fluid and "sterno" cooking fuels. While the growth of methanol's chemical markets is generally on pace with other chemicals at about 3–5% per year, markets for methanol fuels are expanding at a robust 25–40% per annum. In terms of consumer exposure to methanol, the use of methanol transportation fuels—primarily in the act of vehicle refueling—represents the largest potential exposure route. However, the 5-minutes-exposure to the inhalation of methanol vapors from refueling a methanol compatible vehicle is expected to generate less of a methanol uptake than drinking a can of diet soda containing the sweetener aspartame (which metabolizes in the body to methanol).

In the United States, there are now more than 8 million ethanol flexible-fuel vehicles (FFV) on the road although only a few alternative fuel vehicle buffs will recall that FFV technology began with 20,000 methanol/gasoline cars sold in the 1980s and 1990s. The Renewable Fuel Standard, established by the U.S. Congress in 2007, calls for the

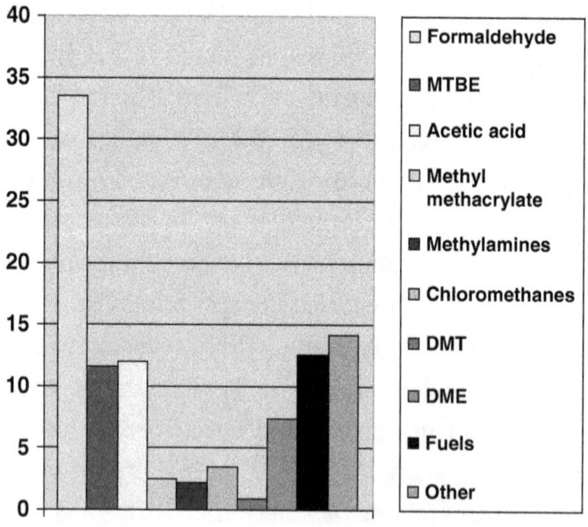

FIGURE 1.2 Global methanol consumption.

use of 36 billion gallons of renewable fuels by 2022, which many translate into a mandate for corn ethanol and cellulosic ethanol. Actually, methanol produced from renewable biomass feedstocks will count too, and may make more sense (and cents).

The State of California often seems like the conscious of the global automotive industry, pushing for the market introduction of more efficient and cleaner vehicle technologies. We can trace this history back to the late 1970s when the California Energy Commission began testing dedicated methanol-fueled vehicles. Operating vehicles on neat methanol had its benefits and drawbacks. These dedicated vehicles would take advantage of methanol's higher octane content (100 octane for methanol versus 87–94 for gasoline) by using higher compression ratios to increase fuel efficiency and dramatically reduce emissions. There were problems with cold-starting vehicle on neat methanol and concerns with the visibility of methanol flames in bright, sunlight conditions. By the early 1980s, the effort turned to methanol FFVs capable of running on a blend of up to 85% methanol and 15% gasoline (called M-85) in the same fuel tank. The use of M-85 assisted with cold starting and imparted visibility to methanol flames. The real drive behind FFV technology was to help overcome the problem of limited availability of methanol fueling stations in the early years of the program. The objective was to introduce large numbers of methanol FFVs, build a broad fueling infrastructure network, then transition back to dedicated methanol vehicles.

With encouragement from the state, a series of initiatives led to the demonstration of 18 different models of methanol-fueled cars from a dozen automakers. The state also established a methanol fuel reserve and entered into 10-year leases with gasoline retailers for the establishment of a network of 60 public retail methanol-fueling pumps and 45 private fleet-accessible fueling facilities. Over 15,000 methanol FFVs would find a home on California's streets and freeways, along with hundreds of methanol-fueled transit and school buses. During the peak of the program in 1993, over 12 million gallons of methanol was used as a transportation fuel in the state. Through these efforts, FFVs were developed as a largely inexpensive "off-the-shelf" technology, and the challenges of dispensing alcohol fuels were solved. In addition, fearing the potential market share loss from growing methanol fuel use, the major oil companies began introducing cleaner "reformulated" gasolines that eroded many of the clean air benefits of using methanol.

Ultimately, only four methanol FFV models moved from prototype demonstration to commercial availability (Ford Taurus 1993–1998 model years; Chrysler Dodge Spirit/Plymouth Acclaim 1993–1994 model years; Chrysler Concorde/Intrepid 1994–1995 model years; and the General Motors Lumina 1991–1993 model years). By the mid-1990s, automakers had already abandoned further development work on methanol, turning instead to work on compressed natural gas and battery electrics. Today, China has picked up the methanol torch, with over 2.3 billion gallons methanol blended in gasoline (M-15, M-30, M-85, and M-100) in 2011 for use in passenger cars, taxis, and bus fleets. Chinese automakers are introducing new models of methanol FFVs, while national fuel standards for methanol fuel blending have been adopted to grow the market in an organized manner. China now views coal-based methanol as a strategic transportation fuel.

This in an important point as the use of methanol as a transportation fuel offers a viable means of transitioning from fossil-based fuels to renewable fuels. Liquid secondary energy carriers have a much bigger market potential than gaseous hydrogen (or liquid hydrogen, t, −253°C) (30). Methanol can be produced from natural gas or coal in the short-term, from biomass in the midterm, and from captured atmospheric CO_2 and renewably generated hydrogen in the long term.

The cost to produce methanol from natural gas is around $0.40 per gallon (Zerbe, 1991), and even discounting for methanol's lower energy content, an equivalent pump price to gasoline for methanol would be 25–50 cents per gallon less than the cost of regular gasoline at the pump. A 2010 study from the Massachusetts Institute of Technology (Cohn, 2010) found that "With deployment of plants using current technology, on an energy-equivalent basis, methanol could be produced from U.S. natural gas at a lower cost than gasoline at current oil prices." This interdisciplinary study went on to recommend that the U.S. government implement an open fuel standard requiring automakers to provide tri-flex-fuel vehicles capable of running on ethanol, methanol, and gasoline.

As a hydrogen carrier, methanol has many advantages for fuel cell vehicles. In 2002, DaimlerChrysler's NECAR 5 completed a cross-country journey from the San Francisco Bridge to the U.S. Capitol Building, achieving a 300-mile range on each tank of methanol fuel. At Georgetown University, five methanol fuel cell buses have been built. Europe's leading motorhome manufacturer, Hymer AG of Germany, has integrated a direct methanol fuel cell to provide autonomous "hotel-load" power for their Hymermobile S-Class. In Japan, Yamaha is now leasing a motorbike powered by a methanol fuel cell. Many of the world's leading consumer electronics companies are preparing to market laptop computers, cell phones, and other devices powered by direct methanol fuel cells. With no tough-to-break carbon-to-carbon bonds, methanol readily releases its hydrogen for fuel cell use.

As a chemical commodity, methanol is a building block for hundreds of products that touch our daily lives. And as an energy resource, methanol is re-emerging as a clear alternative for transportation fuel markets. With abundant feedstocks from natural gas, coal, biomass, and even waste carbon dioxide, methanol has the potential to efficiently and economically serve as the molecular backbone of the global economy.

REFERENCES

CMAI, (2011) World Methanol Consumption, December.

Cohn, D. (2010) The Future of Natural Gas, Massachusetts Institute of Technology, August.

Crocco, J. (2002) The Evolution of the Methanol Industry: From Ancient times to the Future, DeWitt Global Methanol and Clean Fuels Conference, October.

Olah, G. (2006) *Beyond Oil and Gas: The Methanol Economy*, Wiley-VCH.

Olofsson, M. (2012) http://www.sweden.gov.se/sb/d/7534. Last accessed date: Nov. 10, 2012.

Specht, M. and Bandi, A. (1999) "The Methanol-Cycle"—Sustainable Supply of Liquid Fuels, Center of Solar Energy and Hydrogen Research (ZSW).

Williams, R.H., Larson, E.D., Katosky, R.E., and Chen J. (1995) Methanol and hydrogen from biomass for transportation. *Energy Sustain. Dev.* **I** (5), 18–34.

Zerbe, J. (1991) Liquid fuels from wood—ethanol, methanol, diesel. *World Res Rev.* **3** (4), 406–414.

Zuberbühler, U. (2005) Gasification of Biomass—An Overview on Available Technologies, 1st European Summer School on Renewable Motor Fuels, 29–31 August.

2 Methanol: Fate and Transport in the Environment

RULA A. DEEB, TODD L. ANDERSON, MICHAEL C. KAVANAUGH, and LAUREN A. KELL

ARCADIS U.S., Inc., Emeryville, CA, USA

2.1 INTRODUCTION

2.1.1 Release Scenarios

In the United States in 2007, methanol ranked fourth among all chemicals reportedly released by industry to the environment as noted in annual Toxics Release Inventory (TRI) reports required by the U.S. Environmental Protection Agency (USEPA) (USEPA, 2009). These releases were primarily from paper, chemicals, and wood products industries (USEPA, 2009). As shown in Table 2.1, methanol releases from industry in 2006 and 2007 in the United States were primarily to the atmosphere; however, ~15–19% of methanol was directly discharged into groundwater, soil, or surface water during these years. The total reported volume of methanol released to the U.S. environment represents ~1.5% of the total U.S. production volume. In 2001, the United States produced an estimated 3.5–4 million metric tons (mt) of methanol (DeWitt, 2002), with roughly 1.5–2 million of this being "merchant" (for transport/sale) and the remaining 2 million metric tons created and used at the same facility as a feedstock for other products (DeWitt, 2002). Monitoring of methanol in the atmosphere, surface water, or groundwater is generally not required; neither the Clean Air Act (CAA), Clean Water Act (CWA), nor Safe Drinking Water Act

The Toxicology of Methanol, First Edition. Edited by John J. Clary.

TABLE 2.1 Estimated Releases of Methanol in the United States by Industrial Sources

Reported Release to	2006 (million pounds/year)	2007 (million pounds/year)
Atmosphere	145	129
Underground injection	19	13
Land	8	4
Surface water	6	5
Total releases	178	151

Source: USEPA (2009).

(SDWA) includes methanol monitoring requirements. Thus, national monitoring data sets and information on methanol occurrence in air or water are not available (Zogorski et al., 1997).

This chapter evaluates the fate and transport of methanol in soil, groundwater, and surface water in the context of three methanol release scenarios. The three scenarios are as follows:

Scenario 1: Rail Car or Tank Truck Release. Most of the methanol used in North America is imported from overseas. In one estimate, ~7.1 million metric tons was imported in 2006 (PCI-Ockerbloom & Co. Inc., as cited in Alliance Consulting International, 2008). Another estimate puts this value at 5.4 million metric tons as of 2002 (DeWitt, 2002). Once it reaches a port, it must be transported via rail or truck to its final destination. Approximately 1.6 million "merchant" metric tons produced per year in North America (as of 2006) often must be transported to the point of use as well (PCI-Ockerbloom & Co. Inc., 2008). Rail cars and tanker trucks are the two primary land-based methods of inland transportation of methanol (DeWitt, 2002). An accidental release from a rail car or tank truck could take place in a variety of physiogeographical settings, depending on railway and highway alignments, and possibly including environmentally important features such as the desert, the coast, or drinking water sources. A single rail car could release as much as 30,000 gallons

(100 mt) of methanol (DeWitt, 2002; Perry et al., 1984) if fully emptied during such an event. A typical truckload of methanol is 8000 gallons (24 mt) (DeWitt, 2002). To give an example of scale, in 1994, ∼1400 truckloads were delivered throughout California (CEC, 1994).

Scenario 2: Ship or Barge Release. More than 80% of methanol produced in the world is shipped between continents (Alliance Consulting International, 2008). In 2007, ∼40 million metric tons of methanol were consumed worldwide (Alliance Consulting International, 2008). Ocean releases could occur since the vast majority of methanol imported into the United States is by ship, and river releases could occur since barges are used for intergulf or river transport of methanol from regional ports. Typical river barges can carry just under 418,000 gallons, or ∼1255 mt (DeWitt, 2002). Deep-sea transport of methanol increased dramatically during the 1980s and 1990s. Smaller to midsize dedicated methanol ocean-going tankers range in size from ∼20,000 to 50,000 mt; some dedicated methanol tankers have a capacity of ∼100,000 mt (Waterfront Shipping Company Limited, 2010). This corresponds to a volumetric range of oceangoing ship capacities from ∼5 to 32 million gallons (MG). Assuming that methanol imported from Canada to the United States is transported entirely by rail (or truck), and methanol imported from the remaining four primary importing countries (Trinidad, Chile, Venezuela, and Equatorial Guinea) is transported by tanker ship, roughly 1.3 billion gallons (BG) of methanol is imported to the United States each year by ship (DeWitt, 2002), corresponding to perhaps 75 tanker trips.

Scenario 3: Storage/Fueling Facility Release. The third conceptual scenario involves the accidental release of methanol to the environment at a fueling or a storage facility. Methanol is stored at docks and marine terminals in floating roof tanks; these typically have elaborate leak detection and safety systems. Methanol is also stored in aboveground tank farms with aboveground piping and leak detection and fire suppression systems. While spills may occur in these mass storage facilities, the larger concern is the possibility of releases at smaller distribution facilities and from totes and drums (Alliance Consulting International, 2008). As discussed further in Section 2.2, methanol is a commonly used carbon source for denitrification at wastewater treatment plants throughout the United States.

At distribution facilities, methanol underground storage tank (UST) systems are generally similar to gasoline systems, although some differences in materials used may exist (SWRCB, 1999a). New, upgraded UST systems are double-walled and typically have an interstitial leak detection device or other leak detection mechanism. Leak detection depends upon a number of factors, such as the location, volume, and velocity of the leak. Most commonly, leaks occur at the joints or at the dispenser; if a leak occurs at the dispenser it may not be detected. Leak detection systems can be subject to human error because alarms can typically be just turned off without action being taken. Studies have found that while newer USTs are less likely than older, single-walled tanks to leak, even upgraded USTs experienced leaks (SWRCB, 1999b). Whether leaked material enters the environment also depends on the presence of a catch or drip pan. Because of all the potential leak scenarios from a UST, there are large differences in the amount of fluid that may enter environment in this scenario. It is conceivable that methanol releases may occur at methanol fueling facilities at a rate similar to gasoline UST releases. If methanol USTs are located at or near a gasoline dispensing location, subsurface methanol releases may encounter existing gasoline contaminant plumes. Similarly, methanol releases may also encounter chlorinated solvent plumes or other subsurface contamination. In any case, this third scenario represents another important potential route of methanol release to soil and/or groundwater.

2.1.2 Fate in the Environment

Methanol occurs naturally in the environment because of various biological processes in vegetation, microorganisms, and other living species (ENVIRON, 1996). Nevertheless, a large release of methanol to the surface water, soil, or groundwater has the potential to adversely impact the surrounding environment.

TABLE 2.2 Reported Half-Lives of Methanol and Benzene in the Environment

Environmental Medium	Methanol Half-Life (Days)	Benzene Half-Life (Days)
Soil (based on unacclimated grab sample of aerobic/water suspension from groundwater aquifers)	1–7	5–16
Air (based on photooxidation half-life)	3–30	2–21
Surface water (based on unacclimated aqueous aerobic biodegradation)	1–7	5–16
Groundwater (based on unacclimated grab sample of aerobic/water suspension from groundwater aquifers)	1–7	10–720

Source: Adapted from Howard et al. (1991).

Once released into surface waters or the subsurface environment, the fate of methanol depends on numerous environmental factors including: the nature and quantity of the release and physical, chemical, and biological characteristics of the impacted media. Various reports summarize estimates of possible methanol half-lives (the time required for a 50% reduction in concentration) (Table 2.2) in various environmental media. In the atmosphere, methanol would be photooxidized relatively quickly; its reported half-life ranges between 3 and 30 days. In soil or groundwater, rapid biodegradation is expected as well, with reported half-lives ranging from 1 to 7 days. Finally, in surface water following a pure methanol spill, methanol would also be expected to diminish quickly; reported half-lives are between 1 and 7 days as well. In Table 2.2, reported methanol half-lives are compared to those of benzene to illustrate the relatively rapid degradation of methanol.

On the basis of data summarized in Table 2.2, for the plausible release scenarios envisioned, methanol appears unlikely to accumulate in the soil, air, surface water, or groundwater. However, it is unclear whether these reported rates were developed under realistic field conditions or adjusted based on field experience. Consequently, these values represent generic order-of-magnitude half-lives and should be used with caution for any site-specific assessment of the fate and transport of methanol. In Section 2.3, these rates are further evaluated as a function of specific hydrogeologic (subsurface) and limnologic (lake) conditions.

2.2 PARTITIONING OF METHANOL IN THE ENVIRONMENT

2.2.1 Methanol Partitioning Among Environmental Media

The purpose of this section is to describe the equilibrium partitioning of methanol released to the environment among air, water, and soil phases. Partitioning is directly related to the physiochemical properties of methanol, including solubility, vapor pressure, and adsorptivity, and the physical and chemical characteristics of the environmental medium.

When any chemical is released into an environment consisting of more than one medium (e.g., air and water), the chemical tends to distribute itself between these two phases, a process that has been well described by the science of chemical thermodynamics (e.g., Thibodeaux, 1996; Schwarzenbach et al., 1993). In an enclosed system, the concentrations of the chemical in each phase can be predicted assuming that an equilibrium is achieved. In an open system, however, which is undergoing dynamic changes, continual transfer between phases will occur. The rate of these changes is important for predicting the fate and transport of the chemical of interest in both media. For purposes of this analysis, however, the simple equilibrium model provides a basis for assessing the relative distribution of a compound such as methanol between two or more environmental media.

2.2.2 Air/Water Partitioning

The air/water equilibrium partitioning behavior of methanol can be predicted using Henry's Law when methanol concentrations are <100,000 parts per million (ppm) (i.e., <10% by weight). At higher concentrations, nonideal fluid characteristics must be accounted for.

The Henry's Law constant (dimensionless) represents the ratio of a contaminant's concentration in the air to its concentration in the water at equilibrium at a specific temperature and pressure (usually 1 atm). A compound with a Henry's Law constant <0.05 would partition strongly from the gas phase into the water phase when contaminated air is brought into contact with clean water (Zogorski et al., 1997). The value of the Henry's Law constant for methanol given in Table 2.3 is 1.09×10^{-4} [−] at 25°C. Table 2.3 provides this value and other physical and chemical properties of methanol, gasoline, and benzene for comparison. Since Henry's Law constant for methanol is quite low, and much less than the 0.05 noted earlier, atmospheric methanol would partition readily into water droplets. Concomitantly, methanol in water bodies is likely to remain in the aqueous phase. Therefore, once in water, methanol would be difficult to remove by volatilization—either passively or actively (e.g., air stripping).

TABLE 2.3 A Summary of Physical and Chemical Properties of Methanol, Gasoline (BTEX), and Benzene

Physical and Chemical Properties	Methanol	Gasoline	Benzene
Molecular weight (g/mol)	32.04	~100[c]	78.11[c]
Elemental composition by weight			
% Oxygen	50%	(Mix of C4–C14 hydrocarbons)	0%
% Carbon	37.5%		92%
% Hydrogen	12.5%		8%
Specific gravity	0.7915 at 60°F[e]	0.72–0.78[c]	0.88[c]
Boiling point (°C)	64.7	27–225[b]	80.1[c]
Water solubility (mg/l)	Miscible	100–200[c]	1780
Vapor pressure			
(mm Hg) (at 25°C)	126	—	76, 95[c]
(psi) (at 100°F)[d]	4.63	7–15	—
Heat of combustion (kJ/kg)[b]	19,930	43,030	—
Henry's Law constant (atm m^3/g mol)	4.55×10^{-6} 4.42×10^{-6}[c]	—	5.43×10^{-3}[c]

(continued)

TABLE 2.3 (*Continued*)

Physical and Chemical Properties	Methanol	Gasoline	Benzene
Henry's Law constant [–] (at 25°C)	1.087×10^{-4}	—	0.22
Liquid dispersion coefficient (m^2/s) (at 25°C)	1.65×10^{-9}	—	9.95×10^{-10}
Log organic carbon partition coefficient (log K_{oc})	0.921 0.44[c]	—	1.8–1.99 1.50–2.16[c]
Log octanol–water partition coefficient (log K_{ow})	−0.82[f] −0.77[f] −0.75[c] −0.31[c] −0.16[c]	—	2.13 1.56–2.15[c]
Flammability limits (FL)[b]			
Lower (LFL)			—
Volume percent	6.0	1.4	
Temperature (°C)	7	−43	
Upper (UFL)			—
Volume percent	36.5	7.6	
Temperature (°C)	43	−30 to −12	
Flash point (°C)	12[a]	−42	—
Vapor density (at 1 atm; 10°C)[b]	1.4	2–5	3.4

[a]Merck Index (1989).
[b]Machiele (1989).
[c]Zogorski et al. (1997).
[d]USDOE (1991).
[e]Methanex (2006).
[f]Mackay et al. (1992).

2.2.3 Soil/Water Partitioning

The soil/water partition coefficient, K_d, specifies the equilibrium ratio of a contaminant's concentration in/on the solid phase to that in the aqueous phase, thus, determining the distribution of a chemical between soil and water under equilibrium conditions. In addition, the K_d value can be used to estimate the rate of movement of a chemical in groundwater compared to the rate of groundwater flow. For nonionic organic compounds such as methanol, K_d values are a function of the organic carbon fraction of the soil (f_{oc}) and the organic carbon-based partition coefficient (K_{oc} (l/kg)). It can be shown that $K_d = K_{oc} \cdot f_{oc}$ (Schwarzenbach et al., 1993). Values of f_{oc} are site-dependent and are typically low in most subsurface environments (in the order of 0.5%) (Zogorski et al., 1997). K_{oc} values for methanol are low (~8) (Table 2.3) relative to values of other gasoline oxygenates (~12 for MTBE) and aromatics (~100 for benzene). Consequently, in a soil/water environment with typically low soil organic carbon, methanol would be present primarily in the water phase.

For the listed K_{oc} values, it can be shown that retardation of the rate of methanol migration due to adsorption would be negligible (Wood et al., 1990; Zogorski et al., 1997). The retardation factor for methanol and other short-chain alcohols is usually 1. Thus, dissolved methanol would migrate at the velocity of groundwater. For soils with higher levels of organic carbon, some retardation can occur. For example, in soils with an organic carbon fraction >10% (i.e., for $f_{oc} = 0.1$), the K_d is ~0.8, signifying nearly equivalent concentrations of methanol adsorbed on soil and dissolved in water. Consequently, methanol would generally travel through the subsurface at rates significantly higher than constituents in gasoline, including benzene, toluene, ethylbenzene, or xylenes (BTEX). However, owing to rapid dissolution and biodegradation, discussed in further sections, ultimate methanol plume lengths would likely be shorter than BTEX plumes in the event that simultaneous releases had occurred.

2.2.4 Methanol Dissolution

The dissolution of methanol into groundwater from a methanol product spill scenario involves dissolution from a pure methanol pool (M100: pure, 100% methanol). Under this scenario, methanol is likely to leach quickly from the release area into groundwater. For gasoline releases, the aqueous concentrations of gasoline hydrocarbons and ether oxygenates in contact with water can be estimated using Raoult's law (hydrocarbons' solubility in water multiplied by the mole fraction in the gasoline) (e.g., maximum theoretical solubility of MTBE = 48,000 ppm × 11%

by volume in a given gasoline = 5280 ppm) (Poulsen et al., 1992). However, because methanol is completely miscible in water, this method is not applicable, and the concentration must be estimated based on presumed mixing conditions using a methanol/groundwater mixing ratio and a dilution factor.

Assuming that subsurface mixing ratios are between 10 and 100, the maximum range of methanol concentrations would be 0.9–9.1% by volume (7000–70,000 ppm by mass) for an M100 release. However, these high concentrations would only occur in the immediate vicinity of pure methanol releases. Much lower concentrations are anticipated further from the source due to methanol plume dilution and dispersion.

Alternatively, the concentration of methanol in groundwater can be approximated by taking the ratio of the release flow rate and the groundwater velocity and assuming a source area width and vertical mixing ratio. For example, if methanol is released from an underground storage tank at ~0.5 gpm, groundwater is moving at 1 ft/day, the source area width is 1 m, and the vertical mixing ratio is 10–100, the resulting methanol concentration in the groundwater would be between 70 and 700 ppm. These concentrations would be much larger in the immediate vicinity of a catastrophic underground storage tank release (i.e., complete tank failure) or barge release (to surface water). However, in most cases, the infinite solubility of methanol would cause methanol concentrations to rapidly decrease as a result of mixing, dispersion, and diffusion.

Poulsen et al. (1992) evaluated the rate of dissolution of methanol from an M85 fuel (85% methanol and 15% unleaded gasoline) into water. Poulsen et al. (1992) experimentally determined that ~99% of the methanol would leach into the water phase in the time required for three volumes of water, each equal to the initial gasoline volume, to move past the gasoline. They concluded that this would result in a short discrete plume.

The rate of methanol dissolution into the subsurface following a release is similar to the rate of ethanol dissolution because methanol and ethanol have similar properties. Heermann and Powers (1997, 1998) studied the partitioning behavior of ethanol and BTEX compounds from an ethanol-blended gasoline. According to Heermann (1998, personal communication), the aqueous concentration of ethanol due to dissolution of ethanol-blended gasoline is dependent on the concentration of ethanol in the gasoline and on mass transfer limitations in subsurface systems. Heermann showed that when the concentration of ethanol was 10% (by volume) in a gasoline, >99% of the ethanol partitioned into the water phase. Heermann did not quantify the time required for dissolution but determined that it is limited by the rate of ethanol liquid molecular diffusion to the gasoline/groundwater boundary, which may result in longer dissolution times (>1 year).

In conclusion, pure methanol would be expected to quickly dissolve into groundwater and disperse at a rate directly proportional to the local velocity of the groundwater. All of the methanol present in an M85 gasoline would likely be released into the groundwater, but the time for complete dissolution would be site-specific. It is unlikely that either M85 or M100 would exhibit a slow dissolution time (>1 year) because of the large quantities of methanol in contact with the water phase, methanol's infinite solubility in water, and methanol's affinity for the water relative to soil, thus reducing the effects of diffusion-controlled dissolution. However, the time required for dissolution is highly dependent on the source area shape and the types of soils in the subsurface. Therefore, if methanol enters a low-permeability soil horizon with minimal groundwater contact, particularly one with relatively high organic carbon content, dissolution could proceed slowly.

2.2.5 Commingling/Cosolvency Effects

Cosolvency is a chemical phenomenon whereby one chemical dissolved in water increases the aqueous solubility of a second chemical. For example, BTEX compounds tend to be more soluble in concentrated aqueous methanol solutions than in water alone. This preferential solubility is referred to as the cosolubility effect. The possibility of enhanced aqueous solubility of BTEX due to a neat or pure methanol plume that encounters existing subsurface BTEX contamination is an issue for methanol releases from underground storage tanks at existing gasoline stations or otherwise comingled with gasoline/BTEX plumes (Scenario 3) (Donbaldson et al., 1993).

The majority of studies on the effects of methanol cosolvency have focused on increased BTEX solubility within methanol/gasoline mixtures such as M85. Results show that M85 released into the subsurface may produce high concentrations of aqueous methanol as well as elevated aqueous BTEX concentrations near the front of the plume. However, the BTEX distribution in the plume would change as a function of the aqueous/gasoline phase volume ratios (Poulsen et al., 1992; Chen et al., 1997; Beck, 1991). The American Petroleum Institute (API) published a study that evaluated the effect of methanol on the cosolubility of BTEX compounds for M85 (Barker et al., 1991). API concluded that for aqueous methanol concentrations below 8.5% (v/v) (i.e., about a 10-fold dilution factor), BTEX solubility was not increased. As the concentration of methanol increased from 8% to 25%, the BTEX solubility increased linearly with the methanol concentration. Above 25% (v/v) concentration of methanol, the BTEX solubility increased exponentially with the methanol concentration, specifically, to 933 ppm at an equilibrium methanol concentration of 44%. In conclusion, the magnitude of BTEX cosolubility is controlled by the concentration of the methanol plume at the point of BTEX/methanol plume interaction, which is a function of the dilution factor (Barker et al., 1991).

Cosolvency experimentation with M85 was also conducted at the Canadian Forces Base in Borden, Ontario. Similar to previous studies, the Borden study concluded that for methanol concentrations below 0.7% (i.e., below 7000 ppm), there was no distinguishable change in the mobility or solubility of the BTEX components. However, this study concluded that enhanced mobility would likely occur for methanol concentrations >25% (i.e., >250,000 ppm) (Barker et al., 1990; Hubbard et al., 1994). While these concentrations are unlikely in most of a plume area, they can be anticipated in the source area of an M100 or M85 plume with a low dilution factor.

2.3 FATE AND TRANSPORT OF METHANOL IN THE ENVIRONMENT

The chemical and physical properties of methanol are described in Sections 2.1 and 2.2 and summarized in Table 2.3. The following three subsections discuss the fate and transport of methanol following a release, as presented in the three scenarios introduced in Section 1.1, namely: (1) rail car or tank truck spill to soil and/or groundwater; (2) barge/ship spill to surface water; and (3) underground storage tank release to soil and groundwater. Each subsection further describes the sources of methanol into the respective environmental medium and the ensuing loss mechanisms from that medium.

2.3.1 Soil and/or Groundwater Release

Very few data sets are available on the reported levels of methanol in aquatic systems and subsurface environments in the United States. The apparent lack of monitoring for methanol, particularly in aqueous systems, is likely a result of (1) the level of difficulty associated with methanol analysis in the water phase, (2) its short half-life due to its high biodegradation potential, and (3) the lack of governmental monitoring requirements or control mandates specific to methanol for groundwater, surface water, and drinking water (e.g., Clean Water Act, Safe Drinking Water Act).

2.3.1.1 Sources of Methanol in Soil and Groundwater There are several processes related to the use of methanol as a fuel that would be directly responsible for methanol releases to soil and groundwater. The dominant sources are expected to be those discussed in Scenarios 1 and 3—release from a rail car or tank truck and release from an underground methanol storage tank. Precipitation may also serve as a transport mechanism for methanol to groundwater by infiltration through the unsaturated zone. Precipitation, however, has been shown to be a minor source of contaminant transport into the soil and groundwater for other highly water-soluble chemicals, such as MTBE (Zogorski et al., 1997), and is therefore expected to be a minor transport mechanism for methanol into groundwater.

2.3.1.2 Losses of Methanol from Soil and Groundwater As a result of its high solubility, low retardation factor, and ease of biodegradability, the dominant mechanisms of methanol loss from subsurface soil and groundwater are expected to be biodegradation and advection (i.e., transport away from the source area by groundwater flow), with little loss due to adsorption. These two mechanisms are discussed further in the following subsections.

Biodegradation In general, microorganisms have been shown to play a prominent role in governing the fate of fuel hydrocarbons in subsurface aquatic and soil systems (NRC, 1993). When environmental conditions are amenable to microbial activity, aerobic microbial populations oxidize organic contaminants and consume oxygen in the process. Once the localized area has been depleted of the oxygen necessary for aerobic respiration, anaerobic conditions develop and anaerobic biodegradation proceeds. While it is well established in the scientific literature that methanol would biodegrade readily in the subsurface under both aerobic and anaerobic conditions, four factors are required for successful methanol biodegradation: the presence of indigenous methanol-degrading microbial populations, the availability of electron acceptors, the availability of nutrients, and appropriate pH and temperature levels conducive to microorganism growth. These four factors are discussed in the following subsections.

PRESENCE OF INDIGENOUS METHANOL-DEGRADING MICROBES Methanol is widely produced in nature by anaerobic microorganisms responsible for complex aromatic hydrocarbon biodegradation (Heijthuijsen and Hansen, 1990; Oremland et al., 1982). As a result, methanol molecules are ubiquitous in nature. Furthermore, owing to methanol's high solubility in water, methanol molecules are bioavailable to microorganisms that can utilize them as a source of carbon and energy. A wide distribution of methanol-degraders in the environment can therefore be expected (e.g., Brock and Madigan, 1991).

Systematics of methanol oxidation: Methylotrophs, organisms that can grow using only one-carbon compounds such as methanol, are known to use both methane and methanol as their sole carbon and energy source under both aerobic and microaerobic (low oxygen) conditions (Brock and Madigan, 1991). Methylotrophs are of commercial interest in the

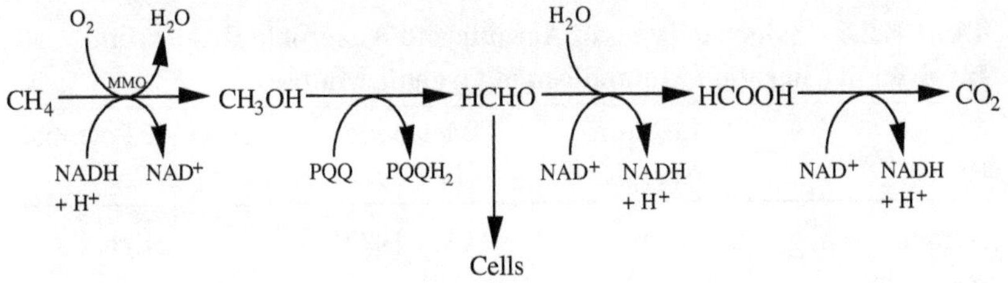

FIGURE 2.1 Methane oxidation pathway of methanotrophic microorganisms. *Source*: Adapted from Chang (1995).

bacterial metabolism of C_1 compounds (Heijthuijsen and Hansen, 1990; Komagata, 1990). Consequently, a large number of methanol-utilizing bacteria have been isolated from a wide variety of natural sources, and most of these isolates have been identified as aerobic, Gram-negative bacteria (Komagata, 1990).

Methanol is also an intermediate of the methane oxidation pathway used by methanotrophic organisms (Figure 2.1); methanotrophs are ubiquitous in nature. Methanotrophs oxidize methane to methanol by the enzyme methane monooxygenase. Methanol is then oxidized to formaldehyde by methanol dehydrogenase (Lehninger et al., 1993; Smeraldi et al., 1994), which is then assimilated into cell material by the activity of either of two pathways, one involving the formation of the amino acid serine and the other proceeding through the synthesis of sugars (Prescott et al., 1996).

Availability of Electron Acceptors Microorganisms obtain energy by transferring electrons from electron donors (in this case, methanol) to electron acceptors. Electron acceptors are compounds that have a lower oxidization state than electron donors; they include molecular oxygen, nitrate, Fe(III), sulfate, and carbon dioxide. The most energetically favored mechanism by which microorganisms oxidize organic compounds is aerobic metabolism (i.e., the use of oxygen as the electron acceptor) (Table 2.4).

The presence of oxygen is a requirement for obligate aerobic microorganisms. Facultative aerobic and anaerobic microorganisms are able to use other electron acceptors when oxygen is not available. Oxygen is

TABLE 2.4 Selected Types of Aerobic and Anaerobic Respiration Involved in Microbial Metabolism of Organic Matter

Process	Electron Acceptor	Metabolic Products	Relative Potential Energy
Aerobic respiration	O_2	CO_2, H_2O	High
Denitrification	NO_3^-	CO_2, N_2	↑
Iron reduction	Fe_3^+	CO_2, Fe_2^+	
Sulfate reduction	SO_4^{2-}	CO_2, H_2S	
Methanogenesis	CO_2	CO_2, CH_4	Low

Source: Taken from Suflita and Sewell (1991).

usually present in and around groundwater recharge areas as a result of infiltrating rainwater. When oxygen is not available, the following electron acceptors are preferentially utilized by microbes in the order shown: nitrate > Fe(III) > sulfate > carbon dioxide (Brock and Madigan, 1991). The redox potential (i.e., available electron acceptors) in subsurface environments is highly site-dependent.

Methanol biodegradation has been shown to take place under both aerobic (oxygen as the electron acceptor) and anaerobic (all electron acceptors besides oxygen) conditions. The aerobic biodegradation of methanol is the most energetically favored mechanism for methanol-degraders and leads to the mineralization of methanol. Microbial growth on methanol with nitrate as the electron acceptor by facultatively anaerobic nitrate-reducing bacteria is the second most energetically favored mode of methanol metabolism (Metcalf and Eddy, Inc., 2003). There are nearly 200 wastewater treatment plants in the United States that currently use methanol as a carbon source to remove nitrate (NO_3^-) from water by anoxic denitrification (Methanol Institute, 2008). Methanol is the most commonly used external carbon source and is dosed at wastewater treatment plants based on the amount of nitrate to be removed as well as dissolved oxygen concentrations. Methanol is used at plants operating a variety of denitritification processes, including tertiary filtration, biological aerated filtration, and moving-bed biofilm reactor processes. Methanol can also be added to activated sludge biological nutrient removal processes, either directly to the anoxic or postanoxic zone(s) depending on plant configuration and operations. Storage of methanol must consider local code requirements because of its flammability (Metcalf and Eddy, 2003; WEF et al., 2005).

Other modes of anaerobic biodegradation of methanol have been studied extensively, resulting in the identification of at least 11 species of methanogenic bacteria that can grow on methanol (Brock and Madigan, 1991).

Other types of anaerobic metabolism include acidogenesis (resulting in the production of acetate or butyrate) (Florencio et al., 1995) and sulfidogenesis (Heijthuijsen and Hansen, 1990). The ability of sulfate-reducing bacteria to oxidize methanol was rarely reported in earlier studies (Esnault et al., 1988; Ward and Winfrey, 1985), but since then evidence has accumulated that suggests that sulfate-reducing bacteria are involved in anaerobic degradation of methanol in sediments, particularly in marine environments (Nanninga and Gottschal, 1986; Esnault et al., 1988). Sulfidogenic methanol-utilizing microorganisms have been shown to oxidize methanol by reducing sulfate to hydrogen sulfide (Heijthuijsen and Hansen, 1990). Finally, some methanol-degrading strains have been shown to utilize hydrogen (Heijthuijsen and Hansen, 1990).

Availability of Nutrients Most microorganisms require nutrients such as nitrogen and phosphorus for cell growth. In general, microbial growth in most subsoils is not limited by nitrogen and phosphorus as long as the contaminant concentrations are in the sub-ppm range (Tiedje, 1993). As stated previously, for M100 releases into the subsurface, methanol concentrations would often exceed 1 ppm, resulting in potentially nutrient-limited conditions. However, as methanol is advected away from the source area, concentrations would decrease and microbial activity would not likely be nutrient-limited.

Appropriate pH and Temperature Levels Most microorganisms grow best in a relatively narrow pH range around neutrality (6–8) (LaGrega et al., 1994). The characteristic range of pH found in groundwater systems (5–9) generally reflects the buffering capacity of the framework carbonate or silicate minerals that make up groundwater aquifers (Chapelle, 1992; King et al., 1992). This range of pH values in aquifers is adequate for the biodegradation of methanol by indigenous subsurface microorganisms since most of the methanol-utilizing bacteria that have been isolated from natural sources have been shown to grow in the neutral pH range (Komagata, 1990). Even if the pH of groundwater were to deviate significantly from the neutral pH range, some strains of methanol-utilizing microorganisms have been shown to prefer acidic pH values ranging from 2.0 to 5.5, while others have been shown to prefer alkaline pH values of 7.0–9.5 (Komagata, 1990).

The temperature of an environment is also an important factor for the growth and activity of microorganisms. Metabolism of subsurface populations accelerates with increased temperatures up to an optimum value at which growth is maximized. Most of the bacteria present in subsurface environments operate most effectively in a temperature range of 20–40°C. This is the range of many natural environments (Chapelle, 1992) since subsurface temperatures within 100 m of the surface are typically within 2°C of the mean annual surface temperature. Temperature should therefore not pose a significant problem for the biodegradation of methanol by indigenous subsurface microorganisms even at temperatures as low as 10°C.

Reports of Methanol Biodegradation Under Various Redox Conditions Simple alcohols such as methanol and ethanol can be degraded in both aerobic and anaerobic environments at rates much faster than gasoline components, such as benzene (Table 2.5). It is important to note in Table 2.5 the difference between laboratory-derived and field-derived degradation rates. Laboratory rates are often higher (an order of magnitude or more) than field-derived rates due to a combination of an elevated laboratory temperature and a larger microbial density in the laboratory growth medium. For example, all the reported benzene rates are laboratory-derived and, thus, likely represent faster degradation rates than would be observed in the field.

Only large concentrations (>100,000 ppm) of alcohols are generally considered to be toxic to most microorganisms and are therefore not

TABLE 2.5 Comparative Rate Constants for Anaerobic and Aerobic Degradation

	Anaerobic Pseudo First-Order Rate Constant		Aerobic Pseudo First-Order Rate Constant
Compound	Electron Acceptor	Range of Rates (day^{-1})	Range of Rates (day^{-1})
Ethanol	NO_3^-	0.53[a,b]	0.23[a,b]
	Fe^{3+}	0.17[a,b]	0.35[a,b]
	SO_4^{-2}	0.1[a,b]	
Methanol	NO_3^-	0.0051–0.88[b,c]	0.019[d,e]
	SO_4^{-2}	0.016–0.34[b,c]	0.011[e,f]
	Anaerobic	0.0059–0.017[b,g]	
	Methanogenic	0.030[b,h], 0.35[b,i]	
Benzene	NO_3^-	0.0095–0.045[b,c]	0.043[b,j]
	Fe^{3+}	0.0065–0.024[b,c]	0.14[b,j]
	SO_4^{-2}	0.0022–0.047[b,c]	
	Methanogenic	0.0039–0.052[b,c]	
	General	0.0062–0.00096[j,k]	

[a]Estimated from Corseuil and Alvarez (1996a) and Corseuil et al. (1996b, 1998).
[b]Determined in a laboratory.
[c]Aronson and Howard (1997).
[d]Hubbard et al. (1994).
[e]Determined from field data.
[f]Barker et al. (1990).
[g]Novak et al. (1985).
[h]Bagley and Gossett (1995).
[i]Florencio et al. (1995).
[j]Rathbun (1998).
[k]Estimated from first principles.

biodegradable (Brusseau, 1993; Hunt et al., 1997a,b). High concentrations of methanol may occur near pure methanol spills and may inhibit microbial activity near the source; however, these concentrations would become more dilute with time and with increasing distance from the source (Katsumata and Kastenberg, 1996).

The biodegradation of methanol has been measured both in the field and in the laboratory studies. In an extensive field study at the CFB Borden site in Canada, the fate and persistence of an M85 fuel was investigated in an aerobic shallow sandy aquifer (Barker et al., 1990). The average concentration of methanol introduced into the aquifer was 7030 ppm. By Day 476 of the experiment, about 99% of the methanol was degraded (Table 2.5 includes the effective first-order degradation rate). Insufficient oxygen was present in the plume to account for the removal of the methanol; the researchers therefore concluded that the methanol degraded first aerobically, then anaerobically (Hubbard et al., 1994).

In a laboratory study, methanol biodegradation was investigated in soils from three different subsurface sites under various redox conditions (Novak et al., 1985). The first site was aerobic and also had high nitrate levels, the second site was anoxic and had low nitrate but high sulfate concentrations, and the third site was anoxic and had low nitrate and sulfate levels. Rapid biodegradation took place in the microcosms at the first site from both the saturated (4.19–4.55 ppm/day) and unsaturated zones (4.44–5.15 ppm/day), although the amount of methanol added was sufficient to create anaerobic conditions. Rapid methanol biodegradation was also observed in the microcosms from the saturated zone of the second site (1.33–3.18 ppm/day) and the third site (1.0 to 2 ppm/day). This study concluded that methanol concentrations up to 1000 ppm were removed in less than a year at temperatures of 10–11°C and at a pH range of 4.5–7.8. Methanol was found to be readily biodegradable in all subsurface soils examined, with biodegradation proceeding more rapidly in the saturated region. This research suggests that methanol contamination in groundwater is unlikely to persist for lengthy periods of time because of its susceptibility to biodegradation under both aerobic and anaerobic conditions.

In another laboratory study, Suflita and Mormile (1993) measured acclimation periods (periods before degradation proceeded) and biodegradation rates of methanol and various other alcohols and fuel oxygenates in anaerobic aquifer slurries at organic concentrations of 50 ppm. These researchers reported a methanol acclimation period of 5 days, the shortest among the alcohols tested (the other alcohols being ethanol, 2-propanol, and *tert*-butanol), as well as an anaerobic methanol biodegradation rate of 7.4 ± 0.7 ppm/day. In a later study, these researchers investigated methanol biodegradation under a range of redox conditions (Mormile et al., 1994). Under nitrate-reducing conditions, methanol was degraded with stoichiometric amounts of nitrate consumed. However, under sulfate-reducing conditions the researchers concluded that the sulfate-reducing bacteria were not utilizing the methanol. Generally speaking, high concentrations of methanol are not used by sulfate-reducing bacteria to an environmentally significant extent. However, sulfate reducers have been found to contribute to methanol depletion in other studies, and at least four species of sulfate-reducing bacteria have been isolated that can utilize methanol (Nanninga and Gottschal, 1986; Esnault et al., 1988).

Other laboratory studies have reported the biodegradation of methanol under various redox conditions and by microorganisms from diverse environments. For example, the methanogenic degradation of methanol was demonstrated for an enrichment culture derived from an anaerobic digester as well as for a pure culture. The rates of methanol degradation by the enrichment and pure cultures without acclimation periods were reported to be 1.92 and 1.37 ppm/min (extrapolated from Figures 2 and 6B, respectively, of Bagley and Gossett, 1995). Fitch et al. (1996) also grew methanol-utilizing cultures in a chemostat culture at methanol concentrations of 1500–2700 ppm and found the cell yield of methanol-grown cells to be consistent with what has previously been reported for methanotrophs. Florencio et al. (1993) investigated the anaerobic biodegradation of methanol at a range of pH values by a culture derived from an anaerobic granular sludge reactor used to treat alcohol distillery wastewater. Florencio et al. (1995) monitored the anaerobic removal of methanol by methanogens and acetogens in upflow anaerobic sludge blanket reactors. Results from these studies showed that methylotrophic methanogens dominated in a broad pH range (4.2–7.2), while significant acetogenesis occurred in the presence of high methanol concentrations. Finally, Beeman and Suflita (1987) measured methanol biodegradation in microcosms from a shallow anoxic aquifer underlying a municipal landfill in Oklahoma. By recovering 70–100% of the methane produced during methanogenesis, they were able to demonstrate that methanogenesis of methanol was occurring.

Adsorption Another methanol loss mechanism from the groundwater is adsorption to soil materials. However, adsorption would not contribute significantly to the loss of methanol from the environment because of the rapid rates of biodegradation in soil and groundwater and vaporization from dry soils. Little adsorption is expected for compounds such as methanol because of its polarity, low K_d value and miscibility, as discussed previously.

Volatilization from Groundwater and Vaporization from Soil Methanol's high solubility and low Henry's constant are expected to result in minimal methanol losses from groundwater because of volatilization or off-gassing. However, in the soil, methanol would vaporize from the residual phase or source area under low soil moisture conditions (ENVIRON, 1996). Vaporization would occur much quicker for M100 fuels than conventional gasoline because of the high vapor pressure of methanol. This may result in migration of methanol away from the source area due to vapor-phase, diffusion-driven transport. Despite the potential for lateral distribution of methanol in the subsurface, analytical soil samples often do not contain detectable levels of methanol near an M85 spill. This is likely due to the mass loss from vapor phase microbial degradation of methanol (Beck, 1991).

2.3.1.3 Methanol/BTEX Commingled Plumes BTEX compounds are the most toxic group of hydrocarbons in gasoline. Naturally occurring biodegradation typically mitigates the persistence of these compounds in the subsurface environment. However, high concentrations of alcohols and ether oxygenates can potentially impede the biodegradation of BTEX (Salanitro, 1993). Alcohols would be biodegraded in preference to BTEX and can therefore consume the oxygen required for BTEX biodegradation. Alcohols can also be toxic or inhibitory to subsurface populations of BTEX-degraders at elevated concentrations (Barker et al., 1990).

Information on the effect of methanol on the biodegradation of BTEX in the field is available from the Borden site study in Canada. Three gasolines (M85, 15% MTBE, and base gasoline) were allowed to migrate in an aquifer (Barker et al., 1990). By Day 476, most of the methanol was degraded, but BTEX compounds in the methanol plume were at higher relative concentrations than in the MTBE and nonoxygenated plumes. These data suggest that methanol may have slowed BTEX biodegradation rates compared to BTEX biodegradation in MTBE-containing and oxygenate-free groundwater. Laboratory experiments found no significant biodegradation of BTEX in the presence of 7000 ppm of methanol but some BTEX biodegradation was noted when the methanol concentration was only 1000 ppm. It was concluded that the mechanism for greater BTEX persistence in the presence of methanol is the initial inhibition of BTEX biodegradation by toxic methanol concentrations followed by inhibition due to the removal of oxygen by preferred methanol biodegradation. Such results are particularly important in assessing the fate of benzene, a known carcinogen (Dean, 1985) and the most recalcitrant monoaromatic under anaerobic conditions.

In another laboratory microcosm study, the biodegradation of BTEX was reportedly impaired at 1.4% (14,000 ppm) methanol and was completely inhibited at higher methanol levels (Novak et al., 1985).

In conclusion, the biodegradation rate of BTEX compounds in a gasoline plume that has become commingled with a methanol plume (Scenario 3) could be significantly retarded in the presence of high methanol concentrations. This reduction in BTEX biodegradation could effectively increase the length of BTEX plumes. A complete evaluation of this effective plume elongation has not been completed for this evaluation, but is likely to be a function of the distance between the methanol leaking UST (LUST) and the gasoline LUST, and of the locations of the two USTs relative to groundwater movement.

2.3.2 Surface Water Release

2.3.2.1 Sources of Methanol in Surface Water The potential dominant source of methanol in surface water is the direct release of neat methanol into a surface water body following accidental release or catastrophic failure during transport, such as is outlined in Scenario 2. Nonpoint sources of methanol include precipitation from the atmosphere, which is not expected to be important, as noted earlier.

2.3.2.2 Losses of Methanol in Surface Water The processes that contribute to methanol losses from surface water bodies include biodegradation, abiotic degradation, volatilization, and bioaccumulation.

As with groundwater, the dominant loss mechanism in surface waters is expected to be biodegradation. Methanol concentrations in the vicinity of a surface water release would rapidly decrease because of advection and dispersion particularly in moving surface waters. Owing to methanol's infinite solubility in water, an M100 release in a surface water body, as discussed in Scenario 2, would disperse to nontoxic levels at a rate much faster than a release of an equivalent volume of gasoline. Aquatic toxicity estimates range from a no-observed-effect-concentration of ~23.75 mg/l (Kaviraj et al., 2004) to an LC50 of ~29,400 mg/l (Material Safety Data Sheet, 2001). The rate of dispersion is directly proportional to the amount of turbulent mixing in the aquatic environment. Tidal flows combined with wind-induced wave action would cause a large methanol spill to rapidly disperse to levels below toxic thresholds. The effect of wave action on the speed of contaminant mixing in surface waters has been measured extensively for gasoline components following release from recreational vehicles (Malcolm Pirnie, 1998). In all cases measured, gasoline components quickly mix throughout the upper layer of the surface water bodies; methanol is even more soluble in water than most gasoline components and would, consequently, mix even more rapidly.

To verify these predictions, several computer simulations were performed to model the advective dispersion of methanol away from the source area (Machiele, 1989). The first hypothetical simulation revealed that a 10,000-ton 3 MG methanol release in the open sea would reach a concentration of 0.36% within an hour of the spill. The second hypothetical simulation of a spill at a rate of 10,000 l/h from a coastal pier exhibited a concentration of <1% at the spill site within 2 hours and to 0.13% within 3 hours after the spill ceased.

Biodegradation The dominant process responsible for removal of methanol in surface water bodies is biodegradation. The reported half-life of methanol in surface waters under aerobic conditions is short and has been reported to be as low as 24 hours (Table 2.2) (Howard et al., 1991).

In flowing water bodies, wind- or current-enhanced mixing typically maintains dissolved oxygen concentrations sufficiently high to support aerobic microbial processes. Even in oxygen-limited environments such as the bottom layers of stratified lakes or reservoirs, anaerobic biodegradation is expected to proceed at rapid rates; the reported half-life for methanol biodegradation under anaerobic conditions ranges from 1 to 5 days (Howard et al., 1991). In addition, the nutrient supply in rivers and lakes is generally not expected to restrict the rate of methanol metabolism because the required nutrient supplies are often constantly recharged by runoff (Alexander, 1994). However, high concentrations of methanol resulting from a large spill in an enclosed area could deplete the surface water of oxygen required to sustain aquatic life.

Abiotic Degradation Abiotic degradation (i.e., nonbiological or chemical) reactions are not likely to contribute significantly to methanol removal from surface water bodies. Hydrolysis reactions usually transform compounds into more polar products; methanol is a very polar molecule and is stable in water. Photolysis involves the transformation of a compound as a consequence of its direct absorption of light. Methanol does not absorb light in the visible spectrum and long-wavelength ultraviolet light (UV). It does absorb very short-wavelength UV, which is present in negligible amounts in solar radiation. In addition, methanol can be naturally oxidized by hydroxyl radicals formed in water by the photolysis of nitrate, nitrite, and hydrogen peroxide resulting from reactions with excited humic materials or from the reaction of H_2O_2 with Fe(II) (Schwarzenbach et al., 1993). Thus, naturally occurring photooxidation of methanol via direct or indirect oxidation is possible; however, these reactions are slow and, thus, not expected to be significant in surface waters.

Bioaccumulation The bioaccumulation of fuel hydrocarbons and certain additives has been studied extensively (e.g., Brault et al., 1994). As shown by various studies, uptake and elimination of organic pollutants in freshwater fish and invertebrates may occur through gill membranes, the efficiency of which depends on blood proteins and the hydrophobicity of the compounds. Bioaccumulation in aquatic organisms is expressed by bioconcentration factors (BCF) (Streit, 1994). To the authors' knowledge, no studies have been reported for methanol

TABLE 2.6 Comparative Values of Octanol–Water Coefficients and Bioconcentration Factors

Compound	Log K_{ow}	Log BCF
Methanol	−0.82[a]	−2.09[a]
	−0.77[a]	
	−0.75[b]	
	−0.31[b]	
	−0.16[b]	
Ethanol	−0.32[a]	−1.63[a]
	−0.31[a]	
	−0.75[b]	
Benzene	2.13[b]	6.8[c]
	1.56–2.15[b]	

[a]Mackay et al. (1992).
[b]Zogorski et al. (1997).
[c]RSP (1994).

bioaccumulation. However, octanol–water coefficients (K_{ow}) and bioconcentration factors for methanol and other oxygenates have been reported (Table 2.6). Octanol–water coefficients may be used as an indicator for bioaccumulation potential and are widely used as a descriptor of hydrophobicity (Mackay et al., 1992). In conclusion, methanol is expected to bioaccumulate slightly less than ethanol, and significantly less than benzene and most other more hydrophobic constituents in gasoline. Small quantities of methanol introduced into mammals as a result of bioaccumulation from methanol fuel releases may be rapidly metabolized, thus reducing any long-term effect.

Volatilization Volatilization is not a primary source of loss for methanol since methanol has a low air/water partition coefficient and the rate of biodegradation most likely exceeds the rate of volatilization. In the environment, methanol has similar properties to ethanol and, therefore, the volatilization rate of ethanol from surface waters can be used to describe the relative volatilization rate of methanol. In a 1998 study, the volatilization rates for several oxygenates and gasoline constituents were calculated based on their chemical properties for various lake conditions (Malcolm Pirnie, 1998; Stocking and Kavanaugh, 2000). Under typical California lake conditions (temperature = 20°C; wind speed >5 mph; epilimnion = ~20 ft), ethanol was predicted to volatilize almost five times slower than BTEX compounds. Under these conditions and ignoring biotic reactions, benzene has a half-life of ~9 days, whereas ethanol has a half-life of ~40 days. Using the model and the conditions discussed earlier, methanol has a predicted half-life of ~60 days. In another evaluation, the half-life of methanol in a river and a pond were modeled to be ~4.8 and 51.7 days, respectively; however, this model included other transport mechanisms besides volatilization (USEPA OPPT, 1994).

In conclusion, despite the slow volatilization and minimal adsorption, methanol would not persist in surface waters because of its intrinsic tendency to be degraded quickly.

2.4 METHANOL ADDITIVES

To date, most evaluations of potential environmental impacts due to releases of methanol have focused primarily on the fate of methanol and circumvented any potential ancillary impacts, for example, those due to additives. Similar to other fuels, methanol may require the use of additives to address performance, air pollution, or health issues.

Several investigations conducted in the 1980s and early 1990s identified the known safety, health, and environmental issues associated with methanol and developed a list of potential additives to address these concerns (Machiele, 1989, 1990). Manufacturers have focused their investigation and testing efforts of additives on luminosity (the ability to see a burning flame), taste (to prevent ingestion), lubricity, and color.

There are several potential additives that may be used in methanol fuel; however, the fate and transport of these additives in the environment is not well understood. This is primarily a result of the proprietary nature of these additives and their complex chemical structure. An evaluation of the fate and transport of potential additives is warranted. In the following sections, two categories of additives, luminosity and taste, are discussed.

TABLE 2.7 Selected Chemical and Physical and Properties of Potential Luminosity Additives

Parameter	Toluene	Hexane
Chemical name	Phenyl methane	*n*-Hexane, dipropyl
Molecular formula	C_7H_8	C_6H_{14}
Molecular weight (g/mol)	92.14	86.18
Specific gravity	0.867	0.6548
Boiling point (°C)	110.6	−95
Log K_{ow} (−)	2.73	4.11
Henry's constant (−)	0.272	0.130
Solubility (in water) (mg/l at 20°C)	526	9.47
Vapor pressure (atm)	26.31	4.89
Appearance	Colorless, liquid with aromatic odor	

2.4.1 Luminosity

Several types of compounds have been identified to increase the luminosity of methanol, including hydrocarbon mixtures, aromatics, aliphatics, olefins, other alcohols, ethers, metal organic salts, and alcohol soluble organic dyes. For example, varying toluene content from 4% to 10% increases methanol fuel's luminosity 10–100 times. Selected chemical and physical properties of two potential luminosity additives are presented in Table 2.7.

In general, the fate and transport of toluene and hexane are well understood based on their physiochemical properties. They have relatively high vapor pressures, are readily biodegraded under aerobic conditions, and exhibit more retardation than methanol because of their lower solubilities and higher soil partition coefficients.

2.4.2 Taste

Studies conducted by Lawless et al. (1983) demonstrated that addition of a bitterant reduces accidental ingestion of a product. The potential health concerns associated with ingestion of methanol-containing fuels have prompted the consideration of bitterants to prevent such ingestion.

TABLE 2.8 Selected Chemical and Physical and Properties of the Bitterant Additive Bitrex

Parameter	Bitrex (Denatonium Benzoate)
Molecular formula	$C_{28}H_{34}N_2O_3$
Molecular weight (g/mol)	446.5
Specific gravity	1.12
Boiling point	UNK
Log K_{ow} (–)	0.9
Henry's constant	UNK
Solubility (water) (mg/l at 20°C)	45,000
Solubility (methanol) (mg/l at 20°C)	690,000
Vapor pressure (atm)	UNK
Appearance	White granules

Note: UNK, Unknown or information not available.
Source: Macfarlan Smith Material Safety Data Sheets.

A possible bitterant candidate is denatonium benzoate, or "Bitrex," which has become a widely used defensive measure to prevent ingestion of products from several industries, including denatured alcohols, medicinal wipes, rubbing alcohol, pesticides, plant nutrients, antifreeze, brake fluid, polishes, cleaners, and disinfectants. Bitrex is also used as a nail-biting deterrent. Preliminary estimates indicate that 10 ppm of Bitrex by volume in M100 would be sufficient to prevent ingestion of methanol. Selected chemical and physical properties for Bitrex are listed in Table 2.8.

According to Material Safety Data Sheets and Bitrex information sheets provided by its Macfarlan Smith distributor, Bitrex is generally stable at neutral pH values for up to two years (Bitrex, 1998). Although no transformation of Bitrex has been observed in acidic conditions, under alkaline conditions it is converted to a hydroxide salt. Chemical or physical variations would not affect the performance of Bitrex since the bitter quality is imparted by the denatonium cation, which remains in solution.

The solubility and octanol–water partition coefficient of Bitrex indicate that it would be less soluble in water than methanol. The fate and transport of Bitrex in the environment, specifically, the biodegradability, retardation, and bioaccumulation are not well understood. However, the manufacturers of Bitrex report a biodegradation half-life, derived under laboratory conditions and temperature of about 20°C, of ~45 days (Bitrex, 1998).

2.5 CONCLUSIONS

Applying the hazard assessment guidelines from USEPA's Office of Pollution Prevention and Toxics, a 1996 study concluded that methanol is not persistent in the environment because it readily degrades in air, soil, and water, and has no persistent degradation intermediates (ENVIRON, 1996). In the event of a catastrophic methanol spill, methanol would rapidly dilute to low concentrations (<1%) and subsequently quickly biodegrade. If cleanup measures are implemented, they must be instituted at a much faster pace compared to petroleum spills in order to capture the methanol plume before significant dilution. However, the relative speed of methanol's biodegradation is expected to result in natural cleanup times that are faster than the active cleanup times for methanol or gasoline releases.

Several additives may be used in methanol. The fate and transport of each of these additives should be further studied.

REFERENCES

Alliance Consulting International. (2008) *Methanol Handling Manual*, Produced for the Methanol Institute, Arlington, VA.

Alexander, M. (1994) *Biodegradation and Bioremediation*, Academic Press, Inc., San Diego, CA.

Aronson, D. and Howard, P.H. (1997) Anaerobic Biodegradation of Organic Chemicals in Groundwater: A Summary of Field and Laboratory Studies. Report prepared for the American Petroleum Institute, Chemical Manufacturer's Association, National Council of the Paper Industry for Air and Stream Improvement, Edison Electric Institute, and American Forest and Paper Association.

Bagley, D.M. and Gossett, J.M. (1995) Chloroform degradation in methanogenic methanol enrichment cultures and by *Methanosarcina barkeri* 227. *Appl. Environ. Microbiol.* **61** (9), 3195–3201.

Barker, J.F., Hubbard, C.E., and Lemon, L.A. (1990) The influence of methanol and MTBE on the fate and persistence of monoaromatic hydrocarbons in groundwater. In Proceedings of the Petroleum Hydrocarbons and Organic Chemicals in Ground Water: Prevention, Detection, and Restoration. pp. 113–127, National Water Well Association, Dublin, OH.

Barker, J.F., Gillham, R.W., Lemon, L., Mayfield, C.I., Poulsen, M., and Sudicky, E.A. (1991) *Chemical Fate and Impact of Oxygenates in Groundwater: Solubility of BTEX from Gasoline-Oxygenate Compounds*, American Petroleum Institute, Publication No. 4531.

Beck, Brian A. (1991) Methanol used as alternative fuel comes with hopes and headaches. *HazMat World*. May, 96–98.

Beeman, R.E. and Suflita, J.M. (1987) Microbial ecology of a shallow unconfined ground water aquifer polluted by municipal landfill leachate. *Microb. Ecol.* **14**, 39–54.

Bitrex (1998) Various Bitrex Product Information Sheets. Macfarlan Smith Limited, Edinburgh, Scotland.

Brault, N., Loranger, S., Courchesne, F., Kennedy, G., and Zayed, J. (1994) Bioaccumulation of manganese by plants: Influence of MMT as a gasoline additive. *Sci. Total Environ.* **153**, 77–84.

Brock, T.D. and Madigan, M.T. (1991) *Biology of Microorganisms*, Prentice Hall, Englewood, NJ.

Brusseau, M.L. (1993) Complex mixtures and water quality. EPA/600/S-93/004, United States Environmental Protection Agency, Washington, D.C.

California Energy Commission (CEC). (1994) Calfuels Plan. Developing an Infrastructure Plan for Alternative Fuel Vehicles. California.

California State Water Resources Control Board (SWRCB). (1999a) *Oxygenate Compatibility and Permeability Report (UST Team 1 Report)*. Advisory Panel on the Leak History of New and Upgraded UST Systems.

California State Water Resources Control Board (SWRCB). (1999b) *Leak Source and Leak Detection Data Collection and Analysis (UST Team 3 Report)*. Advisory Panel on the Leak History of New and Upgraded UST Systems.

Chang, H.L. (1995) Ph.D. Thesis, University of California at Berkeley.

Chapelle, F.H. (1992) *Groundwater Microbiology and Biochemistry*, John Wiley & Sons, New York, NY.

Chen, C.S., Rao, S.C., and Delfino, J.J. (1997) Cosolvent effects on the dissolution of polynuclear aromatic hydrocarbons due to spills of oxygenated fuel in the subsurface environment. *In* Preprints of Extended Abstracts **37**(1)387–389, Proceedings of the 213th ACS National Meeting, San Francisco, CA.

Corseuil, H.X. and Alvarez, P.J. (1996a) Natural bioremediation of aquifer material contaminated with gasoline-methanol mixtures. *Rev. Microbiol.* **27** (1), 19–26.

Corseuil, H., Aires, J., and Alvarez, P. (1996b) Implications of the presence of methanol on intrinsic bioremediation of BTX plumes in Brazil. *Hazard. Waste Hazard. Mater.* **13** (2), 213–221.

Corseuil, H., Hunt, C., dos Santos Ferreira, R., and Alvarez, P. (1998) The influence of the gasoline oxygenate methanol on aerobic and anaerobic BTX biodegradation. *Water Res.* **32** (7), 2065–2072.

Dean, B.J. (1985) Recent findings on the genetic toxicology of benzene, toluene, xylenes and phenols. *Mutat. Res.* **145**, 153–181.

DeWitt & Company Inc. (DeWitt). (2002) Methanol Market Distribution Infrastructure in the United States. Prepared for Methanol Institute. September. Accessed at http://www.methanol.org/pdfFrame.cfm?pdf=DeWittStudyV2.pdf, February 3, 2010.

Donbaldson, C.R., Barker, J.F., and Chatzis, I. (1993) Subsurface Fate and Transport of a Methanol/Gasoline Blend (M85) – A Laboratory Investigation. American Petroleum Institute, Publication No. 4569.

ENVIRON. (1996) Hazard Assessment for Human and Environmental Health: Methanol. Prepared for American Forest and Paper Association, Inc. Washington, D.C.

Esnault, G., Caumette, P., and Garcia, J.L. (1988) Characterization of *Desulfovibrio giganteus* sp. nov., a sulfate-reducing bacterium isolated from a brackish coastal lagoon. *Syst. Appl. Microbiol.* **10**, 147–151.

Fitch, M.W., Speitel, G.E., and Georgiou, G. (1996) Degradation of trichloroethylene by methanol-grown cultures of *Methylosinus trichosporium* OB3b PP358. *Appl. Environ. Microbiol.* **62** (3), 1124–1128.

Florencio, L., Field, J.A., and Lettinga, G. (1995) Substrate competition between methanogens and acetogens during the degradation of methanol in USAB reactors. *Water Res.* **29** (3), 915–922.

Florencio, L., Nozhevnikova, A., van Langerak, A., Stams, A.J.M., and Field, J.A. (1993) Acidophilic degradation of methanol by a methanogenic enrichment culture. *FEMS Microbiol. Lett.* **109** (1), 1–6.

Heermann, S.E. and Powers, S.E. (1997) Modeling cosolvency effects on BTEX partitioning from reformulated gasolines. *In* Preprints of Extended Abstracts **37**(1)389–392, Proceedings of the 213th ACS National Meeting, San Francisco, CA.

Heermann, S.E. and Powers, S.E. (1998) Modeling the partitioning of BTEX in water-reformulated gasoline systems containing ethanol. Accepted for publication in Journal of Contaminant Hydrology.

Heijthuijsen, J.H.F.G. and Hansen, T.A. (1990) C_1-metabolism in anaerobic non-methanogenic bacteria. In G.A. Codd, L., Dijkhuizen and F.R., Tabita, (Eds.), *Autotrophic Microbiology and One-carbon Metabolism*, Kluwer Academic Publishers, Boston, pp. 163–191.

Howard, P.H., Boethling, R.S., Jarvis, W.F., Meylan, W.M., and Michalenko, E.M. (1991) *Handbook of Environmental Degradation Rates*, Lewis Publishers Inc., Chelsea, MI.

Hubbard, C.E., Barker, J.F., O'Hannesin, S.F., Vandergriendt, M., and Gillham, R.W. (1994) Transport and fate of dissolved methanol, methyl-tertiary-butyl-ether, and monoaromatic hydrocarbons in a shallow sand aquifer. American Petroleum Institute Publication No. 4601.

Hunt, C.S., Cronkhite, L.A., Corseuil, H.X., and Alvarez, P.J. (1997b) Effect of methanol on anaerobic toluene degradation in aquifer microcosms. *In* Preprints of Extended Abstracts **37**(1)424-426, Proceedings of the 213th ACS National Meeting, San Francisco, CA.

Hunt, C., Alvarez, P., dos Santos Ferreira, R., and Corseuil, H. (1997a) Effect of methanol on aerobic BTX degradation. In B.C. Alleman and A.L. Leeson (Eds.), *In Situ and Onsite Bioremediation*, Battelle Press, **4** (1), 49–54.

Katsumata, P.T. and Kastenberg, W.E. (1996) Fate and transport of methanol fuel from spills and leaks. *Hazard. Waste Hazard. Mater.* **13** (4), 485–498.

Kaviraj, A., Bhunia, F., and Saha N.C. (2004) Toxicity of Methanol to Fish, Crustacean, Oligochaete Worm, and Aquatic Ecosystem. *Int. J. Toxicol.* **23** (1), 55–63.

King, R.B., Long, G.M., and Sheldon, J.K. (1992) *Practical Environmental Bioremediation*, Lewis Publishers.

Komagata, K. (1990) Systematics of methanol-utilizing bacteria. *FEMS Microbiol. Rev.* **87** (3–4), 291–295.

LaGrega, M.D., Buckingham, P.L., and Evans, J.C. (1994) *Hazardous Waste Management*, McGraw-Hill, Inc.

Lawless, H.T., Hammer, L.D., and Cotina, M.D. (1983) *J. Toxicol. Clin. Toxicol.* **19**, 951.

Lehninger, A.L., Neslon, D. L, and Cox, M.M. (1993) *Principles of Biochemistry*, 2nd edition, Worth Publishers, New York, NY.

Machiele, Paul A. (1989) A Perspective on the Flammability, Toxicity, and Environmental Safety Distinctions Between Methanol and Conventional

Fuels. Prepared for American Institute of Chemical Engineers. U.S. Environmental Protection Agency. Ann Arbor, MI.

Machiele, Paul. (1990) A Summary of the Fire Safety Impacts of Methanol as a Transportation Fuel. SAE Technical Paper Series, SAE International. Warrendale, PA.

Mackay, D., Shiu, W. -Y., and Ma, K. -C. (1992) Illustrated Handbook of Chemical Properties and Environmental Fate for Organic Chemicals. Volume IV: Oxygen, Nitrogen and Sulfur Containing Compounds. Lewis Publishers, Boca Raton, FL.

Malcolm Pirnie Inc., April 3, (1998) Modeling the Volatilization of Methyl Tertiary Butyl Ether (MTBE) from Surface Impoundments. Oakland, California.

Material Safety Data Sheet. (2001) Methyl Alcohol, Reagent ACS, 99.8% (GC). Accessed at http://avogadro.chem.iastate.edu/MSDS/methanol.htm, February 8, 2010.

The Merck Index, Eleventh Edition. (1989) Merck and Co., Inc., Rahway, NJ.

Metcalf and Eddy, Inc. (2003) *Wastewater Engineering: Treatment and Reuse*, 4th edition, McGraw Hill, New York, NY.

Methanex. (2006) Technical Information and Safe Handling Guide for Methanol. Version 3.0. September. Accessed at http://www.methanex.com/products/documents/TISH_english.pdf, February 3, 2010.

Methanol Institute. (2008) Methanol and Wastewater Denitrification. July. Accessed at http://www.methanol.org/pdf/MethanolDenitrification.pdf, March 30, 2010.

Mormile, M.R., Liu, S., and Suflita, J.M. (1994) Anaerobic biodegradation of gasoline oxygenates: Extrapolation of information to multiple sites and redox conditions. *Environ. Sci. Technol.* **28** (9), 1727–1732.

Nanninga, H.J. and Gottschal, J.C. (1986) Isolation of a sulfate-reducing bacterium grown with methanol. *FEMS Microbiol. Ecol.* **38**, 125–130.

National Research Council (NRC). (1993) *In Situ Bioremediation: When Does it Work?* National Academy Press, Washington, DC.

Novak, J.T., Goldsmith, C.D., Benoit, R.E., and O'Brien, J.H. (1985) Biodegradation of methanol and tertiary butyl alcohol in subsurface systems. *Water Sci. Technol.* **17**, 71–85.

Oremland, R.S., Marsh, L.M., and Polcin, S. (1982) Methane production and simultaneous sulphate reduction in anoxic, salt marsh sediments. *Nature* **296**, 143–145.

PCI-Ockerbloom & Co. Inc. (2008) Methanol Executive Overview and Handbook. As cited in Alliance Consulting International, 2008.

Perry, Robert H., Green, Don W., and Maloney, James O. (1984) *Perry's Chemical Engineer's Handbook*, McGraw-Hill Chemical Engineering Series, 6th edition.

Poulsen, M., Lemon, L., and Barker. J.F. (1992) Dissolution of monoaromatic hydrocarbons into groundwater from gasoline-oxygenate mixtures. *Environ. Sci. Technol.* **26** (12), 2483–2489.

Prescott, L.M., Harley, J.P., and Klein, D.A. (1996) *Microbiology*, 3rd edition. Wm. C. Brown Publishers, Dubuque, IA.

Rathbun, R.E. (1998) Transport, Behavior, and Fate of Volatile Organic Compounds in Streams. U.S. Geological Survey Professional Paper 1589. pp. 49-58.

Risk Science Program (RSP). December (1994) Draft Final Report: Intermedia Transfer Factors for Contaminants found at Hazardous Waste Sites Benzene. Office of Scientific Affairs at the California Environmental Protection Agency Department of Toxic Substances Control.

Salanitro, J.P. (1993) The role of bioattenuation in the management of aromatic hydrocarbon plumes in aquifers. *Ground Water Monitor. Remed.* **13**, 150–161.

Schwarzenbach, R.P., Gschwend, P.M. and Imboden, D.M. (1993) *Environmental Organic Chemistry*, John Wiley & Sons, Inc., New York, NY.

Smeraldi, C., Berardi, E., and Porro, D. (1994) Monitoring of peroxisome induction and degradation by flow cytometric analysis of *Hansenula polymorpha* cells grown in methanol and glucose media: Cell volume, refractive index and FITC retention. *Microbiology* **140**, 3161–3166.

Streit, B. (1994) Bioaccumulation processes of organic pollutants by animals of the River Rhine. *Water Sci. Technol.* **29** (3), 145–147.

Stocking, A.J. and Kavanaugh, M. (2000) Modeling Volatilization of MTBE from Standing Surface Waters. *J. Environ. Eng.* ASCE **126** (12), 1131–1136.

Suflita, J.M. and Mormile, M.R. (1993) Anaerobic biodegradation of known and potential gasoline oxygenates in the terrestrial subsurface. *Environ. Sci. Technol.* **27** (6), 976–978.

Suflita, J.M. and Sewell, G.W. (1991) Anaerobic biotransformation of contaminants in the subsurface. EPA/600/M-90/024.

Thibodeaux, Louis J. (1996) *Environmental Chemodynamics: Movement of Chemicals in Air, Water, and Soil. Environmental Chemodynamics*2nd edition, Wiley Interscience.

Tiedje, J.M. (1993) Bioremediation from an ecological perspective. In In Situ Bioremediation: When Does it Work? National Research Council, National Academy Press, Washington, DC.

United States Department of Energy (USDOE). (1991) Assessment of Costs and Benefits of Flexible and Alternative Fuel Use in the US Transportation Sector. Technical Report No. 7: Environmental, Health, and Safety Concerns.

United States Environmental Protection Agency (USEPA). (2009) TRI Explorer, Releases: Chemical Report. Accessed at http://www. epa. gov/ triexplorer/, November 12, 2009.

United States Environmental Protection Agency. Office of Pollution Prevention and Toxics (USEPA OPPT). (1994) Chemical Summary for Methanol. Report No. EPA 749-F-94-013a.

Ward, D.M. and Winfrey, M.R. (1985) Interactions between methanogenic and sulfate-reducing bacteria in sediments, vol. 3, p. 141–179. In Advances in Aquatic Microbiology, Academic Press Inc., London.

Waterfront Shipping Company Limited. (2010) Fleet List. Accessed at http:// www.wfs-cl.com/fleet/fleet-list.htm, February 10, 2010.

Water Environment Federation (WEF) and American Society of Civil Engineers (ASCE)/Environmental and Water Resources Institute (EWRI). (2005) Biological Nutrient Removal (BNR) Operation in Wastewater Treatment Plants. WEF Manual of Practice No. 29; ASCE/EWRI Manuals and Reports on Engineering Practice No. 109. WEF Press, Mc Graw Hill.

Wood, A., Bouchard, D., Brusseau, M., and Rao, P. (1990) Cosolvent effects on sorption and mobility of organic contaminants in soils. *Chemosphere* **21** (4–5), 575–587.

Zogorski, J.S., Morduchowitz, A., Baehr, A.L., Bauman, B.J., Conrad, D.L., Drew, R.T., Korte, N.I., Lapham, W.W., Pankow J.F., and Washington, E.R. (1997) Fuel Oxygenates and Water Quality. Report prepared for the Interagency Assessment of Oxygenated Fuel, National Science and Technology Council, Executive Office of the President, Washington, DC.

3 Human Toxicity

JOHN J. CLARY

Bio Risk, Midland, MI, USA

3.1 INTRODUCTION

The information on human toxicity is primarily from accidentrial or deliberate ingestion of methanol as a substitute for ethanol. The early human data also contains information about occupational over exposure both from the inhalation and dermal routes. Most of the data are a result of a single overexposure. There are a very few reports of methanol toxicity due to repeat sublethal exposure. The difference in the human response to methanol when compared to most of the animal data is due to the difference in metabolism between rodents and humans. In humans, the rate-limiting step in the metabolism of methanol is the buildup of formic acid in the blood due to the saturation of the enzyme, formic dehydrogenase. In rodents, the problem appears to be related to the build up of blood methanol as a result to the saturation of the enzyme catalase, the enzyme, that is, primarily responsible for the breakdown of methanol to formaldehyde (Kavet and Nauss, 1990). In humans, a different enzyme (alcohol dehydrogenase, ADH) carries out the breakdown of methanol to formaldehyde (Kavet and Nauss, 1990).

Humans are more sensitive to the toxic action of methanol than animal species. In the rat, for instance, the acute lethal dose is about 6–10 times higher than the dose, that is, required to produce death in humans (Gilger and Potts, 1955).

This chapter examines dietary exposure to methanol and normal blood levels of methanol and formic acid in humans. The result of

The Toxicology of Methanol, First Edition. Edited by John J. Clary.

humans exposed to large doses of methanol is also discussed. The early history of methanol toxicity is reviewed demonstrating the role of occupational overexposure, both inhalation and dermal, as the major cause of methanol toxicity in humans. Ingestion overdose has been reported more frequently in the later twentieth century and now is the primary cause of methanol toxicity in humans. A section is also devoted to treatment and management of methanol poisoning, but the treatment of an individual with methanol poisoning is an individual issue and should be left to the treating physician.

3.2 EXPOSURE

3.2.1 Dietary

The most important source of background body burdens of methanol is dietary. Methanol is found at low levels in alcoholic beverages. Beer ranges from 6 to 27 mg/l and wine 96 to 329 mg/l while the level of methanol in neutral sprits have been reported to be as high as 1500 mg/l (CERHR, 2003). A Swedish study reported a correlation with blood ethanol and blood methanol following ingestion of alcoholic beverages (Jones and Lowinger, 1988).

Methanol in diet is found in the normal human diet in natural fresh fruits and vegetables, where it occurs as the free alcohol, methyl esters of fatty acids, or as methoxyl groups on pectin (Henze et al., 1953; Francot and Geoffroy, 1956; Meigh, 1957; Sommer, 1962; Casey et al., 1963; Anonymous, 1997). The average concentration of methanol in fruit juices has been estimated to be 140 mg/l of juice with a range of 12–620 mg/l (Francot and Geoffroy, 1956). Eating about 500 g of fruit resulted in methanol concentration in the breath equal to drinking 100 ml of brandy containing 24% ethanol and 0.19% methanol (Taucher et al., 1995).

Aspartame is a dipeptide with a sweetening power of 180–200 times that of table sugar (Cloninger and Baldwin, 1970). It has been widely used in beverages as a sweetening agent. During the absorption and metabolism of aspartame, methanol is released. Methanol makes up 10% of the aspartame molecule. Normal aspartame ingestion has been estimated to be 7.5–8.5 mg/kg per day (for a 70 kg adult that equates to a total methanol intake of 52.5–59.5 mg/day). The upper 99th‰ has been estimated to be 34 mg/kg (238 mg methanol for a 70 kg adult). At doses higher in a control human experiment (100, 150, and 200 mg/kg) aspartame blood methanol concentrations was elevated, reaching its highest-level 2–3 hours after ingestion. Levels return to normal by 8 hours (Stegink et al., 1981).

Methanol in the diet also comes from dimethyl dicarbonate. It is a FDA approved direct food additive, that is, added to wines and fruit juices as a microbial control agent. Hydrolysis yields 48 mg of methanol from each 100 mg of dimethyl dicarbonate added. According to the FDA, an adult human can metabolize 1500 mg of methanol per hour with no adverse effects (FDA, 1988). The FDA has concluded that a tolerable (safe) level of exposure to methanol is 7.1–8.4 mg/kg bw per day (FDA, 1994, 1996).

Formate, the toxic metabolite of methanol in humans, does not accumulate if dietary exposure is below 20 mg methanol per kilogram (Anonymous, 1997).

3.2.2 Environmental

Methanol is used in the industrial production of many important chemicals (formaldehyde, acetic acid, MTBE, glycol methyl ethers, methylamine, and methyl methacrylate). It is also used in many commercial products (paints, varnishes, antifreeze solutions, windshield washer fluids, deicers, and fuels). These different uses may result in low-level environmental or occupational exposure to methanol (Anonymous, 1997).

High concentration of methanol in the air may be slightly irritating to the eyes and mucous membrane (Delbrick et al., 1982; Rowe and McCollister, 1982). Liquid methanol may cause burning or tearing of the eyes. Long-term contact with the skin may result in dry, rough, or chapped skin likely due to the defatting effect of methanol (Delbrick et al., 1982). Other than dry rough skin following repeat exposure, there is little or no information on the effect of chronic low-level dermal exposure to methanol. High level of dermal methanol exposure can result in acute methanol toxicity as discussed in the early occupational cases.

Methanol is readily absorbed through the skin. A worker who had methanol spilled over his legs and did not change his shoes or clothes went blind the next day. After 7 months only transient improvement was noted (MAK, 1991). In another case of dermal exposure, metabolic acidosis was reported in a man, who cleaned a tank with methanol. He wore a positive pressure breathing apparatus, but no protective clothing. He did not change his methanol-saturated clothing for another hour after finishing the job. It was concluded that dermal exposure from clothing was responsible for his acidosis (Downie et al., 1992).

The odor threshold has been reported to be 1500–2000 ppm (Scherberger et al., 1958) while May (1966) reports a higher odor threshold (5900 ppm). In any case, the odor threshold of methanol is much higher than the workplace exposure standard of 200 ppm; and the odor will not warn the individual of exposure until it is much higher than the workplace exposure standard.

The workplace standard, established more than 50 years ago, of 200 ppm appears to be adequate to protect the worker from the effects of airborne methanol for 8 hours work day (MAK, 1958; ACGIH, 1959).

3.3 METABOLISM IN HUMANS

3.3.1 Normal

In humans, methanol is readily absorbed into the body by all routes of exposure. Because, it is water-soluble methanol is found in tissues in direct relation to the tissue water content. Methanol is metabolized to formaldehyde by ADH in humans. Formaldehyde is then broken down to formic acid and then to formate and hydrogen ion by NAD formaldehyde dehydrogenase. Formate forms a complex with terahydrofolate (from dietary folic acid), which is the broken down by formic acid dehydrogenase to carbon dioxide and water. At low levels of methanol, such as what is found in the diet, the breakdown of methanol in humans is rapid and complete (Kavet and Nauss, 1990). See Chapter 7 for more details.

Low levels of methanol are normally found in the saliva, blood, and expired breath in normal individuals, not occupational exposed (Eriksen and Kulkarni, 1963; Larsson, 1965). The range of methanol found in blood of normal individuals (not occupational) exposed ranged from 0.32 to 2.61 mg/l (Sedivec et al., 1981; Lee et al., 1992; Osterloh et al., 1996; Batterman and Franzblau, 1997). The methanol comes from diet and normal metabolism (the breakdown of protein methyl esters).

3.3.2 High Exposure

In the case of high exposure to methanol, the rate-limiting step in the breakdown of methanol in humans is the conversion of formate to carbon dioxide and water. The saturation of formic acid dehydrogenase in humans leads to increase in formate and hydrogen ion in the blood. This buildup results in acidosis and toxicity in humans associated with over exposure to methanol. This response is not seen in rats because of their abilities to clear formate more efficiently than humans (Kavet and Nauss, 1990). The monkey metabolizes methanol much more like humans than rats. When formate clearance in rats and monkeys has been compared, the clearance of formate is 2–21/2 times faster in rats than what is found in monkeys. The half-life in rats for formate clearance is about 12 minutes versus 31 minutes in monkey (Kavet and Nauss, 1990). A comparison of the lethal dose of methanol in humans to other animals shows that the lethal dose in monkey is about three times higher than humans and in the rat about nine times higher than humans. This makes humans more sensitive to the lower dose of methanol than rodents and monkeys. Most toxicity studies in rodents are done at doses above the acute lethal dose in humans. In cases of human fatalities due to methanol poisoning, the formic acid in the blood is estimated to be about 640–1100 mg/l (Wallage and Watterson, 2008).

3.3.3 Over Exposure

3.3.3.1 Symptoms Methanol poisoning in humans starts with signs of intoxication such as central nervous system (CNS) effects (depression, headache, dizziness, nausea, and vomiting) followed by 12–24 hours or more of little or no symptoms then visual disturbances (blurred vision, indistinct vision, altered visual field, and eye pain followed by blindness, abdominal pain, and possible death) (Cook et al., 1991). The latency period, before visual effects are noted, following ingestion of methanol is not an estimate of the severity of the poisoning. The concurrent ingestion of ethanol may help delay the onset of acidosis and other effects of methanol toxicity. Blurred vision strongly supports the diagnosis of methanol toxicity (Barceloux et al., 2002).

Mortality correlates best with the degree of acidosis and formate concentration rather than serum methanol concentration (Barceloux et al., 2002).

Ophthalmoscopic examination following methanol poisoning showed local increase in blood flow of the optic disc was the first change seen in the eyes. This was followed by a white striated edema appearing to surround the retina. The degree of white striated edema appears to correlate with the seriousness of the effect of methanol on vision (Kavet and Nauss, 1990). CNS effects appear when blood methanol levels reach 200 mg/l and blindness is seen above 500 mg/l (Anonymous, 1997).

There are numerous case reports on accidental methanol poisoning. In addition to effects reported above other effects such as long-term paralysis of vocal cords, visual effects, and difficult in walking (Riegel and Wolf, 1966). Magnetic resonance imaging (MRI) and cat (CT) scan effects have also been reported. Focal necroses in the brain white matter and in the Putamen are commonly seen on MRI and CT scans of seriously effected individuals. Bilateral hemorrhagic in the putamen is also seen (Vira-Cartrodeza et al., 2007; Keles et al., 2007; Gaul et al., 1996; Phang et al., 1988; Bessell-Browne and Bynevelt, 2007).

3.3.3.2 Blood and Urine Methanol In addition to clinical symptoms, methanol and its metabolites have been measured in blood urine and breath. Urinary methanol levels greater than 10 mg/l suggests total exposure to methanol was too high (above the TLV) (Ferry et al., 1980). The CERHR (2003) established a "safe level" of methanol in the blood at 10 mg/l. Only at air levels above the workplace exposure limit of 200 ppm does blood methanol exceed this level. In humans exposed by inhalation to methanol at 400 ppm, blood levels of methanol were reported at 13.4 mg/l (Franzblau et al., 1995).

In a study of volunteers exposed to 800 ppm of methanol for up to 8 hours (0.5, 1, 2, and 8 hours), methanol was measured in blood, urine, and breath. Methanol levels were approximately linerly related to exposure concentration, but showed a large interindividual variation. The blood methanol level after 8 hours was 31 mg/l (Batterman et al., 1998).

In a study of volunteers exposed to 400 ppm of methanol for 8 hours, methanol and formate were measured in urine. Methanol levels were not proportional to duration of exposure because of metabolism and other elimination processes. Formate in the urine was not judged to be a useful indicator of air borne methanol exposure because it was poorly related to airborne methanol exposure levels (Franzblau et al., 1997).

3.3.3.3 Urinary Formic Acid Urinary levels of methanol accurately reflect concentrations in the blood over a wide range of concentrations, but urinary excretion of formic acid was too variable to be of value in exposure assessment (Ferry et al., 1980).

3.3.3.4 Breath—Methanol Levels Methanol was measured in the blood and the breath in a study of volunteers exposed to inhalation concentrations of methanol up to 800 ppm for 8 hours with or without exercise. Eight individuals were also exposed dermally by placing the hand in a large beaker of methanol for 16 minutes. Methanol was measured in the blood and the breath before exposure and after 6 and 8 hours of exposure. Significant increase in blood and breath were seen after 6 and 8 hours after exposure to 400 ppm methanol. Breath methanol levels peaked 15 minutes after dermal exposure. On the basis of blood methanol levels, 8 minutes of dermal exposure (one hand in a beaker of methanol) was equal to the blood methanol level seen by inhalation at the TLV (200 ppm) (Franzblau et al., 1995). In a follow-up study, the methanol dermal absorption rates averaged 8.1 +/− 3.7 $mg/cm^2/h$ in individuals, who had exposed their hand in a large beaker of methanol for 16 minutes. Methanol blood levels peaked 0.5 hour after exposure (Batterman and Franzblau, 1997).

The controlled human studies discussed above indicate that the workplace standard of 200 ppm is adequate to prevent clinical symptoms and significant elevation of methanol and its metabolite in the body. Methanol in the blood is below 10 mg/l, a safe level, following exposure at 200 ppm methanol in the air.

3.4 HISTORY OF HUMAN TOXICITY

3.4.1 Occupational

In 1856, a mention was made in the literature that methanol affected the eyes in cabinet marker, hatters, and metal workers. This appears to be the first reference in toxicity of methanol used in an occupational setting (MacFarland, 1855–1856).

In addition to methanol ingestion poisoning, workers who coated the inside of beer vats and tanks reported methanol intoxication. Methanol was used as a solvent, and death and blindness were reported. Both inhalation and dermal exposure were possible source of intoxication (McCord, 1931). In one case, a worker, who was varnishing a vat for 2 days, became intoxicated because of inhalation and dermal contact with methanol. On day 3, he reported headache, unsteady gait, nausea, vomiting, and inebriation. Changes in vision were still being observed 6–7 months later (Tyson, 1912).

In 1906, numerous deaths among methanol users in industry resulted in the passage of legislation, which allows tax-free ethanol for industry uses as a replacement for methanol (McCord, 1931).

Methanol occupational exposure was rare until it started to be used as a solvent about 100 years ago. Occupational exposure in the early 1900s was to dermal and inhalation of the solvent "methanol" and blindness and deaths were reported in workers. By the 1920s increased awareness to the effects of high-level dermal and inhalation exposure to methanol were well known. Methanol toxicity to dermal and inhalation occupation exposure is now a minor concern (Delbrick et al., 1982).

3.4.2 Ingestion

In about 1896, commercial preparation of methanol that was deodorized and purified were made. It was sold as Columbia sprits, eagle sprits, lion d' Or and it was added to products such as witch hazel, Jamaica ginger, vanilla extract, and perfumes. In 1904, nearly 2 million workers used methanol in some form. This additional exposure resulted in the increase in the ingestion of methanol in place of ethanol. In a review, Wood and Buller (1904) reported 54 cases of blindness and death because of drinking or inhalation exposure and an additional 84 cases of death, but no blindness. In addition, 90 cases of blindness, followed by death because of drinking of methylated spirits were reported in this review (Wood and Buller, 1904). By 1913, hundreds of cases of methanol poisoning from ingestion or inhalation exposure had been reported in a review by Baskerville (1913).

Early industrial cases of acute methanol poisoning, via both the dermal and inhalation routes, were reported when methanol was used as a solvent, but reports of oral methanol poisoning rarely occurred in the literature until after 1890 when pure methanol was produced at a low cost. In 1910, some cheap wines, brandies, and whiskey were reported to contained 24–43% methanol as a substitute for the more expensive ethanol (Ziegler, 1921).

It was not until 1923 the poisoning of a group of dockworkers in Hamburg following the ingestion of methanol, that the role of methanol ingestion was first accepted as the casual agent in a mass poisonings (Reif, 1923).

Woods (1912) postulated that on the basis of his experience with methanol poisoning patents, if 10 people drank 4 ounces (~100 grams) of Colombia sprits four would die, two out of the four would develop blindness before they die. Six would live, but two would develop blindness over time (Wood, 1912). On the basis of this postulation, the dose that is lethal to about half of the exposed individuals is about 1.4 g/kg. This value is in line with other estimates that 0.3–1.0 g/kg is a minimal lethal in humans (Kavet and Nauss, 1990). It is difficult to get better or more accurate values based on the actual amount of methanol ingestion, because the value is subjective in most cases. There also appears to a variable susceptibility in the human population (Kavet and Nauss, 1990). The ingestion of ethanol with methanol, which slows methanol metabolism and the hepatic folate status, which is important in the metabolism of formate, both can affect the methanol susceptibility in humans (Anonymous, 1997).

Of the potential exposure routes, ingestion of methanol is by far the largest current concern and most of the recent reports of methanol toxicity in humans are due to ingestion of methanol either accidentally or deliberately (Delbrick et al., 1982).

There are many reports of more recent mass methanol poisoning because of accidental ingestion. Some of the specific reports are discussed below to give an overview of outcome of these poisoning. In one study, only 50 out of 200 people who report drinking the same amount of methanol became ill and 12 died (Uhthoff, 1915). There appears to be a wide range of susceptibility in the human population to the effects of ingestion of methanol. One also has to question the accuracy of the ingestion data reported by the different individuals.

In one incident, 15 ml of a 40% solution of methanol was fatal to one individual, but another person survived 500 ml of 40% methanol. Von Burg (1994) also states that 80–150 ml of 80 proof methanol is usually fatal. No references or data were presented (Von Burg, 1994). There appears to be a great deal of variation in people's response to methanol.

In a review of acute methanol poisoning due to adulterated whiskey was published in 1953 (Bennett et al., 1953). The review contained a good history of 323 cases, clinical history, symptoms, laboratory finding, and potential treatment. Speculation as to the mode of action and earlier cases was discussed. One clear observation in the 323 cases was a latent period of general 24–48 hours following the ingestion of methanol to symptoms. Visual effects may be noted earlier in some patients. Of changes related to the metabolism of methanol, formic acid was noted in the blood and urine of methanol-poisoned individuals. This observation, more than 50 years ago, appears to be a key in the mechanism of methanol poisoning. The role of ethanol on the toxicity of methanol was also discussed, especially its effect on the rate of breakdown of methanol and the buildup of formic acid in cases of poisoning. Acidosis was reported to be the most striking metabolic disturbance in human poisoned with methanol. Symptoms attributed to methanol poisoning included visual disturbances. It fact it was seen to some degree in all 115 patients that reported acidosis out of the 323 cases reported in this review, others who had normal serum bicarbonate levels also reported vision problems. Many reported no vision problems after they recovered from the acute methanol poisoning. A total of 41 patients died, CNS effects were reported including headache and dizziness. Nausea, vomiting, ophthalmic changes, and dyspnea were also reported. In this 1953 report, sodium bicarbonate or sodium lactate was given IV, and the results were dramatic with a rapid return of plasma bicarbonate levels returning to normal with a return to consciousness and lethargy and confusion disappearing. The visual symptoms were also relieved in most cases. The potential role of ethanol treatment in methanol poisoning was also discussed, as an addition to bicarbonate treatment.

Autopsies of fatal cases of methanol toxicity showed cerebral edema with meningeal and subarachnniod petechiae, congestion of lungs, epicardial hemorrhages, and general congestion of abdominal viscera. The eyes from fatal cases showed degeneration of ganglion cells (Bennett et al., 1953).

In 1962, a report of four cases of methanol poisoning were reported following the drinking of duplicating fluid, which contained 40–60% methanol (also 35–55% ethanol). One individual who drank 430 ml of this fluid (50% methanol) died a day and half later. The three other individuals drank a reported 215, 145, and 70 ml of fluid and reported symptoms such as nausea, cramps, vomiting, headache, and blurred vision. All recovered (Kaplan, 1962). It was interesting that the high level of ethanol found in this duplicating fluid could have modified the severity of the methanol effects.

In a study of four cases, who drank neat methanol, one women who drank ~50 ml followed by about 60 ml of neat methanol a day later died. The other three who were reported to have drank 100 ml of neat methanol all survived (Erlanson et al., 1965).

In a single outbreak of methanol poisoning, 97 people were hospitalized of which 28 died (7.5 hours to 12 days after ingestion). The latent period between consumption of methanol and the observations of symptoms (CNS, shock, blurring of vision were 7.5–60 hours. Methanol levels in the blood ranged from 158 to 420 mg/l. In the patients that died shrinkage and degeneration of neurons in the panetal cortex were seen in most cases, putamental degeneration and necrosis, hemorrhage, sponginess were noted in optic chiasma. Severe renal tubular degeneration and necrosis were seen in all cases (Mittal et al., 1991).

In a recent large outbreak of methanol poisoning in Estonia, 154 people were reported to have consumed methanol. A total of 66 individuals died including 43 that did not try to get treatment. Eighteen of the survivors suffered from impaired vision and three had permanent brain damage. Treatment in five community hospitals was limited to ethanol, bicarbonate, and ventilation (some by hand) and dialysis in tertiary care centers (Paasma et al., 2007). In a six-year follow up to this study, in addition to the 21 patients that were reported initially to have permanent damage, new neurological effects were found in another eight of these patients. Vision complications were noted in eight other former patients that showed no permanent effects of methanol poisoning 6 years earlier. Many former patients had died (Paasma et al., 2009). This outbreak illustrates that methanol poisoning of large populations can still be a problem where medical care is limited.

In an individual found dead after consuming methanol, the kidney had the highest level of methanol (5230 mg/kg) followed by the liver, vitreous humor, heart, pericardial fluid, and blood (2840 mg/l) (Pla et al., 1991).

In a study of four people, who ingested methanol copying fluid, three reported nausea, abdominal pain, hypokalemia acidosis, and two had ocular findings (they had the highest serum formate concentrations) 12–24 hours later. Hemodialysis was used as a treatment if blood methanol levels were elevated. The authors concluded that increased blood formate was correlated with clinical signs, but blood methanol was not (Osterloh et al., 1986).

There are other reports of mass methanol poisoning in the literature with similar results. A good recent review of some of these studies is found in American Academy of Clinical Toxicity Practice Guidelines on the Treatment of Methanol Poisoning (Barceloux et al., 2002).

The American Association of Poison Control Center reports on average over 2000 potential cases of methanol intoxication in the recent years. About 10% (~200 individuals) had major clinical signs of methanol toxicity and mortality was seen in about 0.5% (~10 individuals). A few of the deaths are considered suicides (Litovitz et al., 2001).

In a study in Ontario covering 6 years, 43 fatalities were reported due to methanol ingestion. A total of 49% (21) were reported as suicide and the remaining 51% (22) were classified as accidental. In the accidentrial cases, 64% (14) drank methanol as a substitute for ethanol (Liu et al., 1999).

3.4.3 Dermal

Liquid methanol may cause burning or tearing of the eyes. Long-term contact with the skin may result in dry, rough, or chapped skin likely because of the defatting effect of methanol (Delbrick et al., 1982).

Other than dry rough skin following repeat exposure there is little or no information on the effect of chronic low-level dermal exposure to methanol. High level of methanol exposure can result in acute methanol poisoning as discussed in the early occupational cases.

Methanol is readily absorbed through the skin. A worker who had methanol spilled over his legs and did not change his shoes or clothes went blind the next day. After 7 months only transient improvement was noted in his blindness (MAK, 1991). In another case of dermal exposure, metabolic acidosis was reported in a man, who cleaned a storage tank with methanol. He wore a positive pressure breathing apparatus, but no protective clothing. He did not change his methanol-saturated clothing for another hour after finishing the job supporting that dermal exposure from clothing was responsible for his acidosis (Downie et al., 1992).

In Argentina, a home remedy for gastrointestinal complaints in children was to wrap the body in compress soaked in alcohol for several hours. Of 55 children studied, 29 were treated with ethanol, 24 with methanol, and two were treated with both. Two of the ethanol treated children died and most of the other treated ethanol children exhibit ethanol poisoning. Severe acidosis was noted in the methanol treated children (age 1.5 months to 4 years) and CNS depression was noted. A total of 12 of the methanol treated children died from cardiac or respiratory failure between 2 and 10 days following treatment. In addition, three cases of erythema and two cases of flaking skin were noted. The survivors showed no lasting effects (Gimenez et al., 1968).

Methanol has biphasic absorption through the skin (Baranowska-Dutkiewicz, 1980). In an *in vitro* study of methanol skin penetration comparing various species including human (cadaver) the penetration was reached a plateau within 2–4 hours. Permeability was low for swine, medium for rabbit, guinea pigs and the monkey, and high for humans and nude mice (Duprat et al., 1979).

There is no data that suggest that methanol causes allergenic effects. Methanol was not a sensitizer when tested in guinea pigs (MAK, 1991).

3.5 CONTROLLED HUMAN STUDIES

In a controlled inhalation study, normal subjects were exposed to 200 ppm of methanol for 2 two-hour exposures. Urinary methanol levels were reported to exceed 10 mg/l (Osamus et al., 1974).

In a random double blind study with 26 subjects (who served as their own control) who were exposed to 200 ppm methanol or water vapor for 4 hours to evaluate potential neurobehavioral effects of methanol. Exposure to methanol increased serum concentration and urinary excretions of methanol, but did not affect serum formate levels, visual, neuropsychological, or neurobehavioral tests. No significant treatment-related effects were reported (Chuwers et al., 1995).

Potential neurobehavioral effects of methanol were evaluated in 12 human volunteers exposed by inhalation for 75 minutes as sham controls or at the TLV of 200 ppm ($\sim$250 mg/m^3) methanol. The following endpoints were examined before, during, and after exposure to methanol and sham vapors: blood and urinary methanol; plasma formate; oral temperature; blood pressure; subjective mood, alertness, fatigue, workload, and symptom scales; spectral analysis of the electroencephalogram; visual- and auditory-event-related potentials; contingent negative variation; respiration; cardiac interbeat interval; symbol digit substitution task; three choice reaction time; Stroop color-word test; simple reaction time; visual function; critical flicker fusion frequency; hand steadiness; visual search task; Gamberale reaction time task; visual tracking task; Sternberg memory task; interval production task; and speeded addition task.

Methanol exposure produced significant increases in blood and urine methanol concentration. No changes in plasma formate were observed. Most of the neurobehavioral endpoints were unaffected by exposure to methanol; however, statistically significant effects and trends were found for a cluster of variables, including the latency of the P200 component of event-related potentials, performance on the Sternberg memory task, and subjective measures of fatigue and concentration. The changes in neurobehavioral effects were small and did not exceed the normal range. The effects were judged not to be related to methanol exposure at 200 ppm in this limited study (Cook et al., 1991).

In a later study with 26 subjects (random double blind study with subjects who served as their own control) were exposed to 200 ppm methanol or water vapor for 4 hours to evaluate potential changes in urinary and serum formate levels as a result of methanol exposure at the TLV. Urinary and serum formate levels as well as folic acid levels were measured at various times during exposure. No difference was found in serum formate at any time between control and exposed individuals. Only a slight nonsignificant increase in urinary formate was noted at 4 hours. Serum and urinary formate levels were found to be of no value as a biological marker of methanol exposure up to 200 ppm methanol (D'Alessandro et al., 1994).

In what appeared to be the same study, 26 subjects (random double blind study with subjects who served as their own control) were exposed to 200 ppm methanol or water vapor for 4 hours to evaluate potential neurobehavioral effects of methanol. Exposure to methanol increased serum concentration and urinary excretions of methanol, but did not affect serum formate, visual, neuropsychological, or neurobehavioral tests. No significant treatment-related effects were reported (Chuwers et al., 1995).

In a study by the same group methanol serum and urinary levels of methanol were evaluated in a random double blind study with 22 subjects exposed to 200 ppm methanol for 4 hours. Methanol was rapidly absorbed by inhalation (absorption rate constant $= 0.87 \pm 0.6$ hours^{-1}). At the end of the exposure period serum methanol concentration rose from the beginning level of 0.9 ± 0.6 to 6.5 ± 2.7 mg/l as did urinary methanol excretion, but serum formate levels (11.2 mg/l) were not affected. The methanol elimination half-life was 3.2 ± 2.3 hours (Osterloh et al., 1996).

In a controlled inhalation study six individuals (aged 29–55) were exposed to 200 ppm for 6 hours. Blood and urinary levels of methanol and formate were measured before and after exposure and with and without exercise. Blood levels of methanol increased from 1.8 mg/l before exposure to 7–8 mg/l after exposure, but formate levels (~9.6 mg/l) were not affected by methanol exposure at 200 ppm with or without exercise (Lee et al., 1992).

In a study of human volunteers exposed to methanol by inhalation, a time-weighted average of 176 ppm for 7 hours per day for 5 days, urinary methanol and formic acid levels were measured before and after each exposure. Urinary methanol levels were elevated at the end of exposure but had return to baseline before the next exposure. No changes were seen in urinary formic acid levels supporting the lack of accumulation of urinary formic acid at levels close to the maximum allowed workplace exposure levels (200 ppm) (Franzblau et al., 1993).

3.6 IN UTERO EXPOSURE

There is very little information about methanol poisoning and pregnancy. One case reported a 26-year-old woman ingested 250–500 ml of methanol during the 38th week of pregnancy. Maternal serum methanol (230 mg/l) and formate (33.6 mg/l) were elevated initially and mild acidosis was reported. Fetal monitoring showed no signs of fetal distress. The patient was treated with ethanol infusion, bicarbonate, and three treatments of hemodialysis. Delivery took place 6 days after the methanol poisoning, and blood methanol levels were normal. There were no complications or long-term problems in 10 years of follow up in the mother or child (Hantson et al., 1997).

A case of transplacental poisoning was reported in a 28-year-old mother about 30 weeks pregnant. The woman had a HIV infection and asthma and was admitted to an emergency room with respiratory distress and acidosis. Her son was delivered by C-section. Acidosis in the son persisted in spite of fluid, bicarbonate, and blood administration. The infant died on day 4. The mother died on day 10 in spite of ethanol, Fomepizole, fluid, bicarbonate, and hemodialysis treatment. The mother's blood methanol level was 540 mg/l 36 hours after admission to hospital. Her son's methanol level on day 3 was 616 mg/l (Belson and Morgan, 2004).

The two cases discussed above are the only literature reference to the potential effect of methanol on the human fetus. Both were acute poisonings. There is no question that methanol crosses the placenta and death of the fetus can result. There is no data in humans about repeated methanol exposure and possible effects on the fetus. Rodent studies report teratogenic effects following high-methanol exposure during the gestational period. The doses necessary to produce these effects in rodents are above the acute lethal dose in humans. See Chapter 3 for more details.

3.7 REPEAT INHALATION EXPOSURE

There is very little recent literature about the effect of chronic exposure to methanol in the workplace and adverse effects. Symptoms similar to the effects seen following the ingestion of high levels of methanol are seen following exposures above the TLV. In a report of 33 workers, who were engaged in the production of pocket methanol fuel dimmed vision and nasal irritation were occasionally reported following methanol exposure level up to 5500 ppm. Methanol levels in the urine at end of shift correlated with methanol vapor levels (Kawai et al., 1991).

In a study of 66 teacher aides, who were compared to a group of 66 teachers, both groups were exposure to methanol during the use of duplicators in schools. Symptoms such as blurred vision, headaches, nausea, and eyes irritation were reported in 45% of the aides compared to 24% of the teachers. The onset of symptoms increased with increased time spent using the duplicating machines. Some reported skin problems that could have been the result of handling paper still wet with methanol. The study covered 12 schools and 21 duplicator machines. The airborne concentration of methanol was reported to range from 365 to 3080 ppm in the area of the duplicating machines. Fifteen of the areas where the duplicating machines were located exceeded the 15 minutes NIOSH recommended short-term ceiling of 800 ppm. Improvement in ventilation reduced the airborne level of methanol and the incidence of symptoms (Frederick et al., 1984).

In two Russian studies (Chao, 1959; Ubaydullayev, 1963), slight CNS and ocular effects were reported at very low levels (12 mg/m^3). There are many details lacking in these earlier papers and the results are questionable in light of lack of any methanol-related effects reported in later more detailed papers, including controlled human chamber studies at levels of 200 ppm methanol.

It appears that if workplace exposure to methanol is kept below the workplace standard (TLV) of 200 ppm no adverse effects would be expected from repeat inhalation exposure.

3.8 MANAGEMENT OF METHANOL POISONING

Treatment of methanol poisoning is based on many factors and some general guidance is given below. Figure 3.1 gives excellent guidance in the treatment of methanol poisoning but each case should be handled by the treating physician based on the individual laboratory values, the time elapsed following ingestion, clinical symptoms, and so on. The reader is directed to the American Academy of Clinical Toxicology Practice Guidelines on the Treatment of Methanol Poisoning published in 2002 for more detail discussion of dosing regimens and other factors that may be significant in the treatment of methanol poisoning (Barceloux et al., 2002). Included is an excellent comparison of the issues involved in the use of ethanol versus Fomepizole to inhibit or slow down methanol initial metabolism to formaldehyde. Hemodialysis is recommended, if blood methanol levels exceed 50 mg/dl (Kavet and Nauss, 1990).

For patients with blood pH below 7.3 and an indication of methanol ingestion, IV sodium bicarbonate is recommended. Metabolism of methanol should be reduced by treatment with ethanol (oral or IV) or Fomepizole.

Ethanol has approximately 10 times greater affinity for ADH than does methanol and therefore ethanol competitively inhibits or slows

Criteria for diagnosis
Serum methanol concentration > 200 mg/l
History of drinking methanol and Arterial pH <7.3) Serum bicarbonate <20 meg/l (mol/l) Osmolal gap >10 mOsm H_2O)

Treatment

If needed

pH <7.25–7.30
Visual signs/symptoms
Deteriorating vital signs despite intensified supportive care
Renal failure
Significant electrolyte disturbance unresponsive to conventional care
Serum methanol concentration > 50 mg/l

Recommended treatment

Supportive care (folic acid or folinic acid)
Hemodialysis to correct metabolic abnormalities if necessary
Adjust pH 7.3 with bicarbonate if necessary
Reduce serum methanol Fomepizole 15 mg/kg followed by 10 mg/kg every 12 hours for 2 days then 15 mg/kg every 12 hours, continue until serum methanol <20 mg/kg or Ethanol by infusion to maintain serum level of 100–150 mg ethanol/l, monitor ethanol every 1–2 hours, continue until serum methanol <20 mg/kg

For more details and how to combine treatments see Barceloux et al. (2002).

FIGURE 3.1 Management of suspected methanol poisoning.

down the metabolism of methanol (Peterson, 1981). The presence of ethanol reduces the amount of ADH available to oxidize methanol to formaldehyde and on to formic acid (the toxic metabolite). By reducing the production of formic acid, acidosis is prevented or reduced and the toxic effects of methanol lessen. The drug, Fomepizole, is a potent inhibitor of ADH and it therefore also inhibits the metabolism of methanol. Both ethanol or Fomepizole are used in cases of methanol poisoning to slow down the metabolism of methanol and help prevent the resulting build up of formic acid, the toxic metabolite of methanol than cause acidosis, nerve damage, vision effects, and even death. Formic acid has also been shown to inhibit cytochrome oxidase, which is thought to lead to ocular toxicity. The magnitude of formic acid accumulation correlates with the acidosis and the decrease of plasma bicarbonate. Lactic acid also increases as formic acid interferes with intracellular respiration. The increase of these two acids in the body leads to anion gap increase that is seen in methanol poisoning.

Ethanol treatment is more difficult to maintain the proper loading during treatment. Fomepizole presents less problems during treatment but is much more costly and may not always be available (Barceloux et al., 2002).

Formic acid metabolism, which can have a long half life (20 hours), should be enhanced by folic acid or IV folinic acid. Hemodialysis may be used to enhance methanol and formate elimination and to correct metabolic abnormalities (Barceloux et al., 2002).

3.9 CONCLUSIONS

In humans, the most common cause of methanol toxicity have been reported following single overdosing either by accidental or deliberate ingestion of methanol as a substitute for ethanol. This is still the main cause of concern for methanol and humans health. Methanol poisoning in humans starts with signs of intoxication such as CNS effects, depression, headache, dizziness, nausea, and vomiting followed by 12–24 hours or more of little or no symptoms then visual disturbances including blindness, abdominal pain, and possible death. In the case of high exposure to methanol, an increase in formate and hydrogen ion in the blood results in acidosis, which is thought to be the cause of the adverse effects including blindness and death.

In the early years, occupational exposure also resulted in methanol poisoning because of both high inhalation and dermal contact with methanol. It has been demonstrated in controlled human studies at the workplace standard (TLV) of 200 ppm that methanol can safely be used in industrial applications. Workplace practice (prevention of significant dermal exposure) and controlling inhalation methanol exposure below 200 ppm appears to prevent any toxicity, either acute or chronic.

Treatment of methanol poisoning following a single over exposure by the oral, dermal, or inhalation route is usually successful in preventing blindness and death if started as soon as possible after the exposure.

REFERENCES

American Conference of Governmental Hygienists (ACGIH) (1959) Threshold limit values (TLV). *AMA Arch. Ind. Health* **20**, 265.

Anonymous (1997) International Programme on Chemical Safety: Environmental Health Criteria. Methanol 196. WHO, Geneva, 160 p.

Baranowska-Dutkiewicz, B. (1980) Human absorption of industrial poisons through the skin. *Bromatol. Chm. Toksyhol.* **13**, 341–343. (Pol).

Barceloux, D.G., Bond, G.R., Krenzelok, E.P., Cooper, H., and Vale, J.A. (2002) American academy of clinical toxicology practice guidelines on the treatment of methanol poisoning. *Clin. Toxicol.* **40** (4), 415–446.

Baskerville, C. (1913) Wood alcohol: a report of the chemistry, technology and pharmacology of and the legislation pertaining to methyl alcohol. New York State Factory Investigation Commission, vol 2, Appendix 6: 917–1042.

Batterman, S.A. and Franzblau, A.T.S. (1997) Time resolved cutaneous absorption and permeation rates of methanol in human volunteers. *Int. Arch. Occup. Environ. Health* **70**, 341–351.

Batterman, S.A., Franzblau, A.T.S., D'Arcy, J.B., Sargent, N.E., Gross, K.B., and Schreck, R.M. (1998) Breath, urine and blood measurements as a biological exposure indices of short term inhalation exposure to methanol. *Int. Arch. Occup. Environ. Health* **71**, 325–335.

Belson, M. and Morgan, B.W. (2004) Methanol toxicity in a newborn. *J. Toxicol. Clin. Toxicol.* **42**, 673–637.

Bennett, I.L., Cary, F.H., Mitchell, G.J., and Copper, M.N. (1953) Acute methyl alcohol poisoning: a review based on experiences in the outbreak of 323 cases. *Medicine* **32**, 431–463.

Bessell-Browne, R.J. and Bynevelt, M. (2007) Two cases of methanol poisoning: CT and MRI features. *Australas Radiol.* **51** (2), 175–178.

Casey, J.C., Self, R., and Swain, T. (1963) Origin of methanol and dimethyl sulfide from cooked foods. *Nature* **200**, 885.

Center for the Evaluation of Risk to Human Reproduction (CERHR) (2003) NTP-CERHR Expert Panel Report on the Reproductive and Developmental Toxicity of Methanol. NIH Publication No. 03-4478.

Chao, C. (1959) Material on the hygienic standardization of the maximally permissible concentration of methanol vapors in the atmosphere. *Gig. Sant.* **24**, 7–12 (Russ).

Chuwers, P., Osterloh, J.D., Kelly, T., D'Alessandro, A., Quinlan, P., and Becker, C.E. (1995) Neurobehavioral effects of low level methanol vapor exposure in healthy human subjects. *Environ. Res.* **7**, 141–150.

Cloninger, M.R. and Baldwin, R.E. (1970) Aspartylphenylalanine methyl ester: a low calorie sweetener. *Science* **170**, 81–82.

Cook, M.R., Bergman, F.J., Cohen, H.D., Gerkovich, M.M., Graham, C., Harris, R.K., and Siemann, L.G. (1991) Effects of methanol vapors on human neurobehavioral measures. *Health Eff. Inst.* Research Report **42**, 1–45.

D'Alessandro, P., Osterloh, J.D., Chuwers, P., Quinlan, P., Kelly, T.A., and Becker, C.E. (1994) Formate in serum and urine after controlled methanol exposure at the threshold limit value. *Environ. Health Perspect.* **102**, 178–181.

Delbrick, W.R., Kluge, A., and Tauber, U. (1982) The effect of methanol in man and animals. DGMK Berichte Research Report 260-07. Hamburg, Germany.

Downie, A., Khattab, T.M., Malik, M.I.A., and Samara, I.N. (1992) A case of percutaneous industrial methanol toxicity. *Occup. Med.* **42**, 47–49.

Duprat, P., Fabry, J.P., and Gradiski, D. (1979) Cutaneous permeability in vitro: experimental selection of an animal model. *Arch. Mal. Prof. Med. Trav. Secur. Soc.* **40**, 927–938 (Mar. Pollut. Bull. 11).

Eriksen, S.P. and Kulkarni, A.B. (1963) Methanol in normal breath. *Science* **141**, 639–640.

Erlanson, P., Fritz, H., Hagstam, K.E., Liljenberg, B., Tryding, N., and Voight, G. (1965) Severe methanol intoxication. *Acta Med. Scand.* **177**, 393–408.

Ferry, D.G., Temple, T.A., and McQueen, E.G. (1980) Methanol monitoring: comparison of urinary methanol concentrations with formic acid excretion rates as a measure of occupational exposure. *Int. Arch. Occup. Environ. Health* **47**, 155–163.

FDA (1988) Federal Register 53 41325-41329.

FDA (1994) Federal Register 59, 5317-532021.

FDA (2001) Federal Register 61, 26786-26788.

Francot, P. and Geoffroy, P. (1956) Le methanol dans les jus de fruits, les boissons, les alcools e spirttueux. *Rev. Ferment. Ind. Aliment.* **11**, 279–287.

Franzblau, A., Lee, E.W., Schreck, R.M., D'Arcy, J.B., Santrock, J., and Levine, S.P. (1993) Absence of formic acid accumulation in urine following five days of methanol exposure. *Appl. Occup. Environ. Health* **8**, 883–888.

Franzblau, A., Batterman, S.A., D'Arcy, J.B., Sargent, N.E., Gross, K.B., and Schreck, R.M. (1995) Breath, monitoring of inhalation and dermal methanol exposure. *Appl. Occup. Environ. Hyg.* **10**, 833–839.

Franzblau, A., Batterman, S.A., Zhou, N., Stephen, C.J., D'Arcy, J.B., Sargent, N.E., Gross, K.B., and Schreck, R.M. (1997) Elevation of methanol and formate in urine as a biological exposure indices of methanol exposure. *Appl. Occup. Environ. Hyg.* **12**, 367–374.

Frederick, L.J., Schulte, P.A., and Apol, A. (1984) Investigation and control of occupational hazards associated with the use of spirit duplicators. *Am. Ind. Hyg. Assoc. J.* **45**, 51–55.

Gaul, H.P., Wallace, C.J., Auer, R.N., and Chen Fung, T. (1996) MRI finding in methanol intoxification. *Am. J. Neuroradiol.* **16**, 1783–1786.

Gilger, A.P. and Potts, A.M. (1955) Studies on the visual toxicity of methanol. V. The role of acidosis in experimental methanol poisoning. *Am. J. Ophthalmol.* **39**, 63–86.

Gimenez, E.R., Vallejo, N.E., Izurieta, E.M., Roy, E., Rossi, S.M., Lis, M., and Ruiz, B.H. (1968) Percutaneous acute alcoholic intoxication: clinical and experimental studies. *Arch. Arg. Ped.* **66**, 121–135.

Hantson, P., Lambermont, J.Y., and Mahieu, P. (1997) Methanol poisoning during late pregnancy. *J. Toxicol. Clin. Toxicol.* **35**, 187–191.

Henze, R.E., Baker, C.E., and Quackenbush, F.W. (1953) The chemical composition of apple storage volatiles. I. Acids, alcohol, esters. *Proc. Am. Soc. Hortic. Sci.* **61**, 237–245.

Jones, A.W. and Lowinger, H. (1988), Relationship between thee concentration of ethanol and methanol in blood samples from Swedish drinking drivers. *Forensic Sci Int.* **37**, 277–286.

Kaplan, K. (1962) Methyl alcohol poisoning. *Am. J. Med. Sci.* **244;** 170–174.

Kavet, R. and Nauss, K.M. (1990) The toxicology of inhaled methanol vapors. *Crit. Rev. Toxicol.* **21** (1), 22–50.

Kawai, T., Yasugi, T., Mizunuma, K., Horiguchi, S., Hirase, Y., Uchida, Y., and Ikeda, M. (1991) Methanol in urine as a biological indicator of occupational exposure to methanol vapor. *Int. Arch. Occup. Environ. Health* **63**, 311–318.

Keles, G.T., Orguc, S., Toprak, B., Ozaslan, S., and Sakarya, M. (2007) Methanol poisoning with necrosis corpus callosum. *Clin. Toxicol.* **45**, 307–308.

Larsson, B.T. (1965) Gas chromatography of organic volatiles in human breath and saliva. *Acta Chem. Scand.* **19**, 159–164.

Lee, E.W., Terzo, T.S., D'Arcy, J.B., Gross, K.B., and Schreck, R.M. (1992) Lack of blood formate accumulation in humans following exposure to methanol vapors at the current permissible exposure limit of 200 ppm. *Am. Ind. Hyg. Assoc.* **53**, 99–104.

Litovitz, T.L., Klein-Schwartz, W., White, S, Cobaaugh, D.J., Youniss, J., Omslaer, J.C., Drab, A., and Benson, B.E. (2001) 2000 Annual report of the

American Association of Poison Control Centers Toxic Exposure Surveillance System. *Am. J. Emerg. Med.* **19**, 337–95.

Liu, J.J., Daya, M.R., and Mann, N.C. (1999) Methanol related deaths in Ontario. *J. Toxicol. Clin. Toxicol.* **37**, 69–73.

MacFarland, J.F. (1855–1856) The methylated spirit and some of its preparations. *Pharm. J. Trans.* **15**, 310.

MAK (1958) cited in MAK (1991). H. Greim (Ed.), *Occupational Toxicants: Critical Data Evaluation for MAK Values and Classification of Carcinogens*, Vol. **16**, Wiley-Vch, Weinheim.

MAK (1991) *Occupational Toxicants: Critical data evaluation for MAK values and classification of carcinogens.* H. Greim (Ed.), Vol. **16**, Wiley-Vch, Weinheim.

May, J. (1966) Odor thresholds of solvents for evaluating solvent odors in air. *Staub-Reinhalt Luft* **26**, 385–389 (Ger).

McCord, C.P. (1931) Toxicity of methyl alcohol (methanol) and inhalation following skin absorption. *Ind. Eng. Chem.* **23**, 931–936.

Meigh, D.F. (1957) Volatile compounds produced by apples. II. Alcohol and esters. *J. Sci. Food Agric.* **8**, 313–326.

Mittal, B.V., Desai, A.P., and Khade, K.B. (1991) Methyl alcohol poisoning: an autopsy of 28 cases. *J. Postgrad. Med.* **37**, 9–13.

Osamus, T., Kenji, N., and Sumie, F. (1974) Methods of evaluating the exposure to methanol and methyl acetate. *Redo Kazaku* **50**, 239–248 (Japan).

Osterloh, J.D., Pond, S.M., Grady, S., and Becker, C.E. (1986) Serum formate concentrations in methanol intoxication as a criterion for hemodialysis. *Ann. Intern. Med.* **104**, 200–203.

Osterloh, J.D., D'Alessandro, A., Chuwers, P., Mogadeddi, H., and Kelly, T.A. (1996) Serum concentrations of methanol after inhalation at 200 ppm. *J. Occup. Environ. Med.* **38**, 571–5781.

Paasma, R., Hovda, K.E., Tikkerberi, A., and Jacobsen, D. (2007) Methanol mass poising in Estonia: outbreak in 154 patients. *Clin. Toxicol.* **45**, 152–157.

Paasma, R., Hovda, K.E., and Jacobsen, D. (2009), Methanol poisoning and long term sequelae – a six year follow up after a large methanol outbreak, in Estonia: outbreak in 154 patients. *BMC Clin. Toxicol.* **9**, 5.

Peterson, C.D. (1981) Oral ethanol doses in patients with methanol poisoning. *Am. J. Hosp. Pharm.* **9**, 1024–1027.

Phang, P.T., Passerini, L., Berendt, R., and King, E.G. (1988) Brain hemorrahage associated with methanol poisoning. *Crit Care Med.* **16**, 137–140.

Pla, A., Hernandaez, A.F., Gil, F., Garcia-Alonso, M., and Villanueva, E. (1991) A fatal case of the ingestion of methanol: distribution in postmortem tissues and fluids including pericardial fluid and vitreous humor. *Forensic Sci. Int.* **49**, 193–196.

Reif G. (1923) Uber die giftigkeit des methylalkohol. *Deut. Med. Wehnschr.* **49**, 183.

Riegel, H. and Wolf, G. (1966) Severe neurological deficiencies as a consequence of methanol alcohol poisoning. *Fortschr. Neurol. Psychiatr.* **34**, 346–351.

Rowe, V.K. and McCollister, S.B. (1982) Alcohols. In G.D. Clayton and F.E. Clayton (Eds.), *Patty's Industrial Hygiene and Toxicology*, Vol. **2** D, John Wiley & Sons, New York, pp. 4528–4541.

Scherberger, R.F., Happ, G.P., Miller, F.A., and Fassett, D.W. (1958) A dynamic apparatus for preparing air vapor mixture of known concentration. *Am. Ind. Hyg. Assoc. J.* **19**, 494–498.

Sedivec, V. Mraz, M., and Flek, J. (1981) Biological monitoring of persons exposed to methanol vapour. *Int. Arch. Occup. Environ. Health* **48**, 257–271.

Sommer, H. (1962) Uber das physiologische schicksal des aus pektin freignemachten methyl alkohols, Ind, *Obst-Gemueseverwert.* **47**, 172–173.

Stegink, L.D., Brummel, M.C., McMartin, K., Martin-Amat, G., Filer, L.J., Baker, G.L., and Tephly, T.R. (1981) Blood methanol concentration in normal adult subjects administered abuse doses of aspartame. *J. Toxicol. Environ. Health* **7**, 281–290.

Taucher, J., Lagg, A., Hansel, A., Vogel, W., and Lindinger, W. (1995) Methanol in human breath. *Alcoholism Exp. Res.* **19**, 1147–1150.

Tyson, H.H. (1912) Amblyopia from the inhalation of methyl alcohol. *Arch. Opthalmol.* **16**, 459–471.

Ubaydullayev, R. (1963) A study of hygienic properties of methanol as an atmospheric air pollutant, in B.S. Levine (transl). *USSR Literature on Air Pollution and Related Occupational Diseases – a Survey* **17**, 39–45.

Uhthoff, W. (1915) Beitrag zu den Sehatorungen durch methylalkohol. *Klin. Monatabl. f. Augenh.* **54**, 48.

Vira-Cartrodeza, A., Perez-Castrilon, J.L., and Dyuenas-Laitam, A. (2007) Magnetic resonance imaging in methanol poisoning. *Clin. Toxicol.* **45**, 429–430.

Von Burg, R. (1994) Toxicology updates. *Methanol. J. Appl. Toxicol.* **14**, 309–313.

Wallage, H.R. and Watterson, J.H. (2008) Formic acid and methanol concentration in death investigation. *J. Anal. Toxicol.* **32**, 241–247.

Wood, C.A. and Buller, F. (1904) Poisoning by wood alcohol cases of death and blindness from Columbian sprits and other methylated preparations. *J. Am. Med. Assoc.* **43**, 972–977, 1058–1062, 1117–1123, 1213–1221, 1289–1296.

Wood, C.A. (1912) Death and blindness from methyl or wood alcohol poisoning with means of prevention. *J. Am. Med. Assoc.* **59**, 1962.

Ziegler, S.L. (1921) The ocular menace of wood alcohol poisoning. *J. Am. Med. Assoc.* **77**, 1160.

4 General Animal and Aquatic Toxicity

JOHN J. CLARY

Bio Risk, Midland, MI, USA

4.1 INTRODUCTION

The toxicity of methanol has been generally well studied in animals. In addition to the general toxicity in animals, separate chapters in this book address potential reproductive/developmental effects (Chapter 5) and cancer studies (Chapter 8). Animal tests, in general, have been done over the years to obtain predictive information on health effects of chemicals in humans. Investigation of methanol toxicity in animals is somewhat difficult to correlate with human response because of differences in metabolism between animals and human. Methanol is metabolized through the same pathways in humans and animals, but by different enzymes, and different rate of metabolites formation resulting in the differences in methanol-induced toxicity between humans and animals.

There is abundant data on the potential acute health effects of methanol in humans. Most information on the human health effects of methanol is derived from clinical observations following accidental or intentional ingestion.

Because of metabolism difference between animal and humans as well as the nature of exposure in humans, additional chapters on metabolism (Chapter 7) and human toxicity (Chapter 3) are presented

The Toxicology of Methanol, First Edition. Edited by John J. Clary.

in this book to put the animal studies on methanol into some perspective for humans.

As far as animal testing is concerned, a special mention of very large research program on methanol toxicity conducted by the New Energy Development Origination (NEDO) of Japan is warranted. NEDO was evaluating methanol as a potential fuel for power stations. This program was a very large undertaking. Unfortunately, the results of these studies were never published by NEDO. The first phase of the research was made available in English in a 90-page summary (NEDO, 1986). The research in monkeys, rats, and mice were made available in the second English summary of 300 pages (NEDO, 1987). There is a great deal of information provided in these summaries as much as you would find in many published papers and these summaries are the basis of the NEDO data presented below.

The NEDO acute inhalation and subacute inhalation studies in monkeys, as well as the chronic inhalation studies in monkeys, rats, and mice are reviewed in this chapter. Other NEDO studies are discussed in other chapters (Chapters 5, 7, and 8) in this book. The NEDO inhalation carcinogenicity studies in rats and mice are reviewed in Chapter 8 on cancer, while the NEDO intraperitoneal studies on metabolism in rats and monkeys are reviewed in Chapter 7 on metabolism in this book. The NEDO inhalation studies of teratology in rats and the two-generation study in rats are reviewed in Chapter 5 on developmental and reproductive effects. The NEDO studies on environmental fate are presented in Chapter 2.

Studies on toxicity in aquatic species are also presented in this chapter.

4.2 ACUTE TOXICITY

4.2.1 Oral

The most common animal toxicity available on most chemicals is oral acute toxicity (LD_{50}). There are numerous acute oral studies on methanol; some going back more than 75 years. In many of the early studies, the number of animals, sex, clinical signs, and age are not given.

Most of these acute oral studies are not GLP nor done to any standard protocol, but demonstrate low acute oral toxicity. Most of the acute oral data was previous summarized by American Methanol Institute (2001).

Table 4.1 gives a listing of available data on methanol acute oral toxicity.

TABLE 4.1 Mortality from Single Ingestion of Methanol in Various Animal Species

Species	LD_{50}, mg/kg	Comments	References
Rat	5,628		RETC (1994)
Rat	9,100		Welch and Slocum (1943)
Rat	9,540		Weil (1972)
Rat	11,520		Union Carbide (1994)
Rat	13,700		Smyth et al. (1941)
Rat	7,914		BASF (1975)
Rat	12,900		Deichmann and Mergard (1948)
Rat	10,300		Kimura et al. (1971)
Rat	7,300		RTECS (1994)
Rat	5,065		BASF (1961)
Rat	12,750		Cornish and Adefuin (1967)
Rat	6,200–12,900		Lington and Bevan (1994)
Rat	9,000–9,500		Gilger and Potts (1955)
Rat	8,300		Alder et al. (1938)
Rats			
14 day	5,861		Kimura et al. (1971)
Young adults	10,300		
Old adults	6,970		
Mouse	5,800		Chikin and Shadurskil (1969)

(continued)

TABLE 4.1 ***(Continued)***

Species	LD_{50}, mg/kg	Comments	References
Mouse	14,390		Smith and Talyor (1982)
Mouse	7,300–1,000		Lington and Bevan (1994)
Dog	6,300	LD80 100% methanol	Haskell et al. (1921)
Dog	6,300	MLD50% methanol	Leo (1927)
Dog	8,000		WHO (1997)
Dog	7,500		RTECS (1944)
Dog	2,500–9,000	One died 4 gm/kg and One died 9.0 gm/kg Seven given 2.5–8.0 g/kg lived	Gilger and Potts (1955)
Miniature pig	>5,000	CNS effects, no ocular toxicity	Dorman et al. (1993)
Monkey	7,000–9,000	Monkey given 5 or 6 g/kg survived MLD 7000	Cooper and Felig (1961)
Monkey	6,000	Four out of eight monkeys died following treatment with sodium bicarbonate	Potts (1955)
Monkey	2,000–3,000	Acidosis, semicoma, and ophthalmologic	Gilger and Potts (1955)
Rabbits	7,000	100% mortality	
Rabbit	7,200	MLD	Baer (1898)
Rabbits	14,200	LD_{100}	Munch and Schwartze (1925)
Rabbit	14,400	LD_{50}	Munch (1972)
Rabbit	2,100	100% mortality	Gilger and Potts (1955)

Rodents, rabbits, and dogs suffer from ataxia, narcosis, coma, and death after high methanol doses. No acidosis or ophthalmologic changes typically seen in humans at high doses were noted in animals except in monkeys (Gilger and Potts, 1955).

The lethal oral dose of methanol in animals, especially rodents is much higher than the lethal oral methanol dose in humans (~1000 mg/kg bw), highlighting the differences in metabolism between various species (Roe, 1982). The LD_{50} dose in rodents ranges from 5085 to 14,400 mg/kg bw. The lethal methanol dose in monkeys is reported to be as low as 3000 mg/kg (Kavet and Nauss, 1990), which is much closer to the lethal dose of ~1000 mg in humans, while the lethal methanol dose in dogs is reported to be between 4000 and 8000 mg/kg bw and in rabbits the lethal methanol dose is between 2000 and 14,000 mg/kg.

On the basis of the acute oral data in rodents, methanol would be classified (Hazard Category IV) as practically nontoxic or very low toxicity by USEPA (2008).

In addition to acute oral studies on mortality, neurobehavioral effects were evaluated in one study. Male Long-Evans rats were gavaged with single methanol doses of 1000, 2000, and 3000 mg/kg bw representing about 10%, 20%, and 30% of the LD_{50}. Ten minutes after dosing, the animals were subjected to a fixed wheel running test to assess operant running (FR20). The rats displayed no signs of overt intoxication such as gait disturbance, but did exhibit a significant, dose-related reduction in FR20 response at all dose levels suggesting that 1000 mg/kg was a Low Observed Adverse Effect Level (LOAEL) (Youssef et al., 1993).

Monkeys receiving methanol doses higher than 3000 mg/kg by gavage showed ataxia, weakness, and lethargy within a few hours of exposure. These signs tended to disappear within 24 hours and were followed by transient coma in some of the animals and death (Potts, 1955; Potts et al., 1955).

In a screening study, mortality was noted in monkeys given 3000, 4000, 6000, or 8000 mg/kg bw of methanol orally, but not at a dose of 1000 or 2,000 mg/kg bw methanol. Intoxication, acidosis, and ocular effects were reported in the monkeys similar to what had been reported in humans (Gilger and Potts, 1955).

Contrary to observations by Gilger and Potts (1955), there was no evidence of marked acidosis in 12 rhesus monkeys after methanol doses up to 6000 mg/kg bw (20–30% methanol in water). There was no hyperventilation, no increase in urinary excretion of organic acids, or shift in serum bicarbonate or death. Blindness was seen in only one surviving monkey dosed with 9000 mg/kg bw. The effect was judged to be transient 4 days after exposure (Cooper and Felig, 1961).

The metabolism of methanol in monkeys is more like the metabolism in humans and the toxic response seen in monkeys is more like humans than seen in other animal species. At higher oral doses of methanol, the monkeys showed severe intoxication, extended edema in the retina and in the optic papilla, and the pupils were wide and nonresponsive.

Six of eight rhesus monkeys dosed with 6000-mg/kg bw methanol followed by sodium bicarbonate to counteract the acidosis, exhibited cystic degeneration of the outer retinal granular layer, and in one animal there was evidence of significant demyelinisation of the optic nerve. One out of eight monkeys survived for 9 days. Histological lesions were seen in the putamen and nucleus caudatus in three of eight monkeys (Potts et al., 1955).

Four out of the eight rhesus monkeys died following oral dosing of 6 gm/kg and treatment with sodium bicarbonate to counteract the acidosis. This was one of the first studies to show improved survival at lethal does as a result of treatment to counteract the acidosis. Most animals showed retinal edema (confined to perimascular regions in some and extending to the entire fundus in others) (Potts, 1955).

Gilger and Potts (1955) concluded that when compared to the mean lethal oral dose of methanol in humans, the rats had a mean lethal dose that was nine times higher, while the mean lethal dose in the rabbit was seven times higher, in the dog nine times higher, and in the monkey three times higher than the mean lethal dose of methanol in humans.

4.2.2 Dermal

Table 4.2 presents mortality from single dermal treatment of methanol in various animal species.

According to an early report, all rats survived 35,000 mg/kg bw after occlusive skin application of methanol, while deaths were reported at

TABLE 4.2 Mortality from Dermal Exposure to Methanol in Various Animal Species

Species	LD_{50}, mg/kg	Comments	References
Rat	35,000–45,000		Eulner and Gedicke (1955)
Rabbit	15,840		Union Carbide (1994)
Rabbit	15,800		Lington and Bevan (1994)
Rabbit	17,000		Carnegie-Mellon (1981)
Monkey	~1,600	Occluded	McCord (1931)

45,000 mg/kg bw (Eulner and Gedicke, 1955). In rabbits, the dermal LD_{50} of about 15,840–17,000 mg/kg bw has been reported (Carnegie-Mellon, 1981; Union Carbide, 1994).

In rhesus monkeys, four daily doses (occluded) of 400 mg/kg bw methanol (total 1600 mg/kg bw per day) caused sickness within 24 hours, and eventually death. After four daily doses of 1000 mg/kg bw methanol (total 4000 mg/kg bw), severe illness took place followed by death on the second day. Rabbits and rats also exposed to methanol under occluded conditions (doses not given) died. The rabbit appeared to be less susceptible to methanol than the monkey or rat (McCord, 1931).

These dermal studies demonstrate that methanol can be absorbed through the skin in rats, rabbit, and monkey in significant amounts to cause toxicity and death, but based on the rodent data methanol would be classified (Hazard category IV) as practically nontoxic or very low toxicity by USEPA classification for dermal toxicity (USEPA, 2008).

4.2.3 Inhalation

Table 4.3 lists the mortality from single inhalation exposure to methanol in various animal species.

High concentrations of methanol in air are needed to produce lethality in rodents. The LC_{50} of methanol in rats was reported to be between 64,000 and 145,000 ppm for exposures ranging from 8 to 1 hour, respectively. In rodents for inhalation toxicity, methanol would be classified (Hazard category IV) as practically nontoxic or very low toxicity by USEPA (2008).

TABLE 4.3 Mortality from Single Inhalation Exposure to Methanol in Various Animal Species

Species	Length of Exposure	LC_{50} ppm	Comments	References
Rat	1 h	145,000	Head only	DuPont (unknown)
Rat	4 h	64,000		RTECS (1994)
Rat	4 h	73,000		Smith and Taylor (1982)
Rat	4 h	98,600	Runny eyes and nose, labored breathing, staggering, apathy, and narcosis	BASF (1980)
Rat	6 h	67,300		BASF (1980)
Rat	8 h	64,000	Saturated vapors	Smyth (2001)
Rat	8 h	>22,500		Lington and Bevan (1994)
Mouse	6 h	41,000		Scott et al. (1979)
Mouse	6 h	>40,000		Lington and Bevan (1994)
Mouse		30,600–152,800		Mashbitz et al. (1936)
Mouse	134 min	61,100		Von Burg (1994)
Mouse	6 h	41,000		NEDO (1997)
Cat	45 h	65,700		Von Burg (1994)
Cat	6 h	23,600		Von Burg (1994)
Cat		20,000–36,900		Witte (1931)
Monkey	1–18 h	10,000–40,000	Mortality	McCord (1931)

In a very early methanol inhalation studies in rhesus monkeys indicated that lethal concentrations (percentage mortality not reported) at ~10,000 ppm (after 18 hours) and ~40,000 ppm) (after 1–4 hours). Blindness associated with optic atrophy was reported, but eventual recovery was observed (McCord, 1931). The age of this study, the lack of mortality data, and the lack of measured air concentration of methanol (values calculated based on amount of methanol vaporized and air volume) make it difficult to assess what might be the LC_{50} in the monkey.

Narcotic effects were reported in mice exposed to 30,800–1,528,000 ppm methanol from 89 to 153 minutes. Overall 45% of the mice died (Mashbitz et al., 1936). In another study, mice were in a state of narcosis but survived 48,000 ppm methanol for 3.5–4 hours daily for 6 days, but mice exposed to 54,000 ppm methanol died after a cumulative 54 hours of exposure (Wesse, 1928). Rats became drowsy, but survived when exposed to 49,700 ppm methanol for 1 hour (Lehmann and Flury, 1943).

The RD_{50} (the concentration necessary to decrease respiratory frequency by 50%) is considered a measure of pulmonary and sensory irritation. The RD_{50} values of methanol in mice are reported to be 25,300 to 41,524 ppm indicating that methanol is not a potent pulmonary or sensory irritant (Lington and Bevan, 1994).

In an inhalation study in rats to evaluate the effect of methanol on serum hormones lutenizing hormone (LH), follicle stimulating hormone (FSH), prolactin, and testosterone were measured in male rats exposed to 0, 200, 5000, or 10,000 ppm methanol for 6 hours Some were scarified at the end of the 6-hour exposure period and others at 24 hours after exposure. In addition both acclimated and nonacclimated rats were evaluated to access the effect of prior handling on the hormone levels. FSH and testosterone were not affected in rats sacrificed immediately after 6-hour exposure to methanol. LH was affected by prior handling and methanol exposure at different time periods. At the 24-hour sacrifice, the results showed a prior handling affected the LH response more than the methanol exposure. No effects on tissue or body weight were noted. Exposure to 5000 ppm methanol and higher resulted in changes in testosterone, prolactin, and LH levels. The NOAEL was 200 ppm (Cooper et al., 1992). This lack of effects on methanol on these pituitary and testicular hormones support the TLV of 200 ppm in contrast to earlier studies that report adverse effects on testosterone and gonadotropins at 200 ppm immediately after 6 hour exposure (return to normal by 18 hours after exposure) (Cameron et al., 1984, 1985).

4.2.4 Intraperitoneal

The LD_{50} in rats given methanol intraperitoneal (i.p.) was approximately 9.5 g/kg bw and in the mouse given methanol i.p. was 10.5–11.5 g/kg bw (Gilger et al., 1952). These doses are similar to oral LD_{50} in rats and mice.

Four rhesus monkeys, four pigtail monkeys, and rats were given 4 gm/kg intraperitoneal doses of methanol, one out of four rhesus and one out of four pigtail died. The surviving pigtail monkeys had coma and severe acidosis, while only one of the surviving rhesus monkey showed mild acidosis, suggesting a difference in response in different strains of monkeys. The pigtail monkey is much more sensitize and more like humans than the rhesus monkey. The rats showed no adverse effects (Clay et al., 1975).

4.2.5 Subcutaneous

The LD_{50} in mice following subcutaneous injection was about 10 gm/kg bw but in neonatal mice the reported valve was 4 gm/kg bw (Goldenthal, 1971).

4.2.6 Intravenous

An intravenous (IV) dose of 4.2 gm/kg resulted in death in a rabbit shortly after treatment (Gilger and Potts, 1955).

In dogs given 2.5–4.7 gm/kg methanol (IV), transient ocular symptoms, similar to the response seen in monkeys given methanol orally were reported (Marc-Aurele and Scheiner, 1960).

4.2.7 Other Acute Studies

There are several short-term studies in rats receiving high doses of methanol ($>$1000 mg/kg). Doses $>$1000 mg/kg are reported to be lethal in humans and above the level were catalase (CAT), the enzyme in rodents that breaks down methanol to formaldehyde, is saturated (Horton et al., 1992). These studies in rats exposed at levels $>$1000 mg/kg looked at oxidative stress (when the antioxidant capability of the target cell becomes insufficient to prevent the increased formations of reactive oxygen species (ROS)). The antioxidant status was evaluated by the increase in lipid peroxidation and a decrease of enzymatic and non-enzymatic antioxidants.

In one study, male Wister rats received 1.5 gm/kg bw methanol by gavage. Groups of six rats were sacrificed at 6, 12, 24, 48, 120, and 168 hours after dosing and antioxidant enzyme activity as well as ascorbate, α-topcopherol, nonprotein, and protein bound sulfhydryl compounds levels were determined in liver, erythrocytes, and serum. In addition, blood pH, lipid peroxidation (thiobarbituric acid reacting substances (TBA), and liver cytotoxicity was also measured. In the liver superoxide dismutasc (SOD) did not change, while (CAT) activity in the liver increased at 6 and 12 hours and then returned to normal. Glutathione peroxidase (GSH-Px) was significantly decreased at 24 and 48 hours, while glutathione reductase (GSsG -R) was decreased 6–24 hours in each group. Nonprotein and protein bound sulfhydryl compounds were reduced at 12–120 hours and ascorbate concentration was decreased only at 12 hours. Lipid peroxidation (TBA) was increased from 12 to 48 hours in the liver after oral dosing.

In the red blood cells, antioxidant enzyme level and ascorbate, α-topcopherol, nonprotein, and protein bound sulfhydryl compounds levels were decreased except for CAT, which was increased at 6 and 12 hours. TBA reactive substances were increased from 6 to 120 hours.

In the serum, SOD was increased during the first 24 hours and the other antioxidant enzymes and ascorbate, α-topcopherol, nonprotein, and protein bound sulfhydryl compounds levels were decreased. TBA reactive substances were increased up to 120 hours.

Blood pH was increased at 6 hours after dosing and then decreased falling below control level by 48 hours. Serum aspartate amino-transferase and alanine aminotransferase levels were also increased indicating liver damage.

This study demonstrates that in addition to damaging the liver in rats, that methanol impairs antioxidant defenses in the liver, erythrocyte, and serum following a single oral dose of 1.5 gm/kg bw of methanol (Skrzydlewska and Farbiszewski, 1996).

In a later study of similar design, male Wistar rats received 3 gm/kg bw of methanol. Rats were scarified at 6, 14, 24, 48, 120, and 168 hours after dosing and antioxidant enzyme activity as well as ascorbate, α-topcopherol, malonaldehyde (MDA) were determined in liver, erythrocytes, and serum. GSH-Px, GSsG-R, GSH, and ascorbate concentrations were significantly decreased in the liver, erythrocyte, and in the blood serum, while CAT and SOD increased initially and the returned to normal by 48 hours after dosing. SOD and α-topcopherol in the erythrocytes were decreased, while MDA in the liver, erythrocyte, and serum were elevated. This study at twice the dose as the previous study also supports that methanol impairs some enzymatic and nonenzymatic antioxidant defenses in the liver, erythrocyte, and serum following a single oral dose of 3.0 gm/kg bw of methanol (Skrzydlewska and Farbiszewski 1998).

In a short-term study in rats, 6 gm/kg bw of methanol (as a 50% solution in isotonic saline) was given by gavage to groups of six male Wistar rats. A group was sacrificed at 6, 12, 24 hours and 2, 5, or 7 days after treatment. Brain and liver were removed and enzyme assays were conducted for superoxide dismutase (SOD), glutathione peroxidase (GSH-Px), glutathione reductase (GSsG -R), and catalase (CAT) in each group. Glutathione (GSH) and ascorbate concentration were also determined at each time point. Lipid peroxidation was assayed as thiobarbituric acid reacting substances (TBA).

SOD activity was significantly depressed after 2 and 5 days returning close to normal by day 7 in the brain. SOD activity was increased at hour 6, 12, and 24 and then returning close to normal in the liver. (GSH-Px) and (GSSG -R) activity in the brain was initially increased for the first 24 hours then returned to normal, while in the liver (GSH-Px) and (GSSG -R) activity was decreased throughout most of the 7 days after treatment. CAT levels were not significantly affected in the brain during the 7-day period, but increased in the liver reaching a peak at day 5 before returning to normal on day 7.

GSH was decreased in both the brain and liver during most of the 7-day period with the biggest decrease seen in the brain during the day at day 7 and in the liver at 24 hours. Ascorbate decreased in both the brain and liver with a decrease seen in the brain at 12 hours and in the liver at through out the 7 days. TBA was increased in the brain, almost doubling by 7 days with increases seen in the liver for up to 5 days.

These observations support a decrease in antioxidant activity in the brain and liver following very high oral methanol dosing in rats. The oxidative mechanism in the brain and liver appear to be different. In the liver, peroxide is primarily removed by CAT, while GSH-Px is more important in the brain (Skrzydlewska et al., 1998).

In another study, the effect of antioxidants (triolox derivative U-83836E and *N*-acretylcystiene) were evaluated on lipid peroxidation in erythrocytes of methanol-treated rats. Previous work has demonstrated that methanol ingestion in rats increased lipid peroxidation of erythrocytes. Intraperitoneal injects of these antioxidants (triolox derivative U-83836E and *N*-acretylcystiene) in rats receiving 3.0 gm/kg of methanol orally, partially prevent lipid peroxidation in the erythrocytes. These effects may be due to the effect of these added antioxidants on the production of ROS (Dobrzyska et al., 1999).

In a short-term repeat study in rats, the effect of methanol on antioxidant status and lipid peroxidation in the lymphoid organs (lymph glands, spleen, thymus, and bone marrow) in male Wistar rats (180–200 g/bw) was investigated following 2.37 g/kg bw (25% LD_{50}) intraperitoneally for 1, 15, or 30 days. Repeated dose of 3.16 or 4.74 gm/kg bw intraperitoneally for 15 days or more resulted in mortality. Lipid peroxidation was estimated by determining MDA and was elevated at each time period (higher with increase time of treatment) in the lymph nodes, spleen, thymus, and bone marrow.

All enzymatic (superoxide dismutase, CAT, and glutathione peroxidase) and nonenzymatic antioxidant (reduced glutathione and vitamin C) were significantly elevated following 1-day treatment in thymus, spleen, lymph nodes, and bone marrow, but these indices were significantly lower than the control on day 15 and lower still on day 30 also indicate oxidative damage on the immune cells of the lymphoid organs. When free radical increases as a result of methanol treatment, it appears that endogenous antioxidant enzymes increase to remove the free radicals and prevent oxidative damage. When the antioxidant defense mechanism is overwhelmed by free radicals, an increase in lipid peroxidation occurs. These suggest that damage to lymphoid organ cell membrane due to increase in ROS is possible.

Liver and kidney function (serum aspartate aminotransferase, alanine aminotransferase, urea, and creatinine) were significantly increased when compared to controls at day 15 and 30 indicating damage to the liver and kidney (Parthasarathy et al., 2006).

In another study, the increased formation of free radicals in Sprague Dawley rats was detected by electron spin resonance spectroscopy following a single dose of methanol (i.p. 4.5 gm/kg or gavage 7 gm/kg). The same free radical adduct was detected in the bile and urine 2 hours after dosing. The free radical can react with oxygen to form reactive oxygen species (Kadiiska and Mason, 2000).

In still another study, hydroxymethyl radicals were detected rat liver microsomes and nuclei of NADPH-dependent process following treatment with methanol. The results suggest that both rat liver nuclei and microsomes are able to generate free radicals during NADPH-mediated methanol transformation (Castro et al., 2002).

In the acute methanol studies in animals, high levels of methanol were needed to achieve a toxic response by all routes. Methanol would be generally characterized as being practically nontoxic or very low toxicity in rodents following a single oral, dermal, or inhalation exposure. Rabbits and dogs give similar response to methanol exhibiting similar symptoms at a lower lethal dose than rodents. The lethal dose in humans is several folds lower than the lethal dose in animals making it obvious that the human is more sensitive to the acute effects of methanol than animals.

4.3 IRRITATION

4.3.1 Dermal

Methanol caused minor to moderate skin irritation in rabbits. Methanol (0.1 ml) tested under OECD guidelines # 405 using six New Zealand White albino rabbits produced a Draize score of 3.72 at 72 hours after treatment (Jacobs, 1990). Moderate skin irritation was observed in rabbits when 500 mg of methanol was applied under occluded conditions for 24 hours. The reaction was probably the result of defatting of the skin seen with alcohols in general. In another study, methanol (amount not given) was nonirritating when applied under occluded conditions for 30 hours (MAK, 1991).

4.3.2 Eye

Methanol caused minor to moderate eye irritation in rabbits (Carpenter and Smyth, 1946).

In three other studies conducted under OECD guideline, methanol was slightly to moderately irritating to rabbit eye. In another test, no eye irritation was recorded in rabbit eye treated with 50 mg (MAK, 1991).

4.4 SENSITIZATION

Methanol was not a sensitizer when tested in modified Magnusson–Kligman maximization guinea pig test (BASF, 1979; MAK, 1991).

4.5 REPEAT EXPOSURE—INHALATION

4.5.1 Non-Human Primates

In a pilot study to set doses for a larger subacute monkey inhalation study, monkeys (*Macaca fascicularis*) were exposed to methanol by inhalation at 0, 3000, 5000, 7000, or 10,000 ppm (reduced to 5000–6000 ppm at night) for 21 hours per day (NEDO, 1986, 1987). Monkeys were exposed for 6 days at 7000 and 10,000 ppm, 14 days at 5000 ppm, and 21 days at 3000 ppm. The design of this study is somewhat unusual for an inhalation study with nearly continuous exposure rather than the normal 6 hour per day for 5 days per week used in most inhalation studies. This design offers much less recovery time per day than the normal inhalation study and is used in all NEDO study where nearly continuous exposure was the condition desired by the experimental design.

One female monkey was exposed per level (except two monkeys were used at 5000 ppm). Parameters evaluated were effects on vision, clinical signs, hematology, histopathology, clinical chemistry, and survival. The results demonstrated that the monkeys tolerated 3000 ppm methanol after a few days, based on physiological and clinical signs, but higher levels of methanol exposure resulted in reduced movement, weak knees, vomiting, and dyspnea. The two monkeys exposed at 5000 ppm were sacrificed or died on day 14 or 15. Monkeys exposed at higher level were also sacrificed. Body weight was decreased in monkey exposed to 10,000 ppm. Monkeys exposed to 5000 ppm methanol and higher had reduced blood pH (acidosis), decreased feed consumption, body temperature, increased lipid levels, and nervous system effects. Blood methanol was 5250 mg/l and blood formic acid was 1210 mg/l in the 5000 ppm monkey. The lower dose animals had blood methanol levels of approximately 80 mg/l and blood formic acid levels of 30 mg/l. No effects on clinical chemistry, but slight changes in the central nervous system (hyperplasia of reactive astrocytes in the basal ganglion) were noted in animal exposed at 3000 ppm methanol. Fatty degeneration of liver, degeneration of basal ganglion, and other nervous system effects were seen in a dose-related manner at monkey exposed at 3000 ppm and higher, but no effects on lung, thyroid, trachea, GI tract, pancreas, spleen, urinary tract, and female organs. In the monkeys exposed at 7000 ppm methanol and higher, ECG showed cardiovascular effects, increase in alkaline phosphatase, and an increase in white blood cells. Body weight decrease was observed in monkeys exposed at 10,000 ppm methanol. No effect on water consumption, protein, GOT, GPT, A/G ratio of urobilinogen, and bilirubin was noted. No treatment-related effects were noted in retina, optic nerve, corpus geniculatum, and calcarinus. The NOAEL was 3000 ppm and the LOAEL was 5000 ppm.

This study is a follow-up to the earlier pilot inhalation study in monkeys conducted by NEDO (1986, 1987). In this study, monkeys were exposed by inhalation to 1000, 2000, 3000, or 5000 ppm methanol for 21 hours per day (NEDO, 1986; NEDO, 1987). This study had no control group. Three female monkeys were exposed at each level. Monkeys were exposed to methanol for 12 days at 5000 ppm, 20 days at

2000 and 3000 ppm and 7 months at 1000 ppm. Recovery time (no exposure) was 1 and 4 months for monkeys exposed at 5000 ppm, 6 months for monkeys exposed at 3000 ppm, and 12 months for monkeys exposed at 2000 ppm, and 1 and 6 months for monkeys exposed at 1000 ppm. Parameters evaluated effects on vision, body weight, clinical signs, hematology, histopathology clinical chemistry, and survival.

The results showed that the monkeys tolerated 1000, 2000, and 3000 ppm methanol, but 5000 ppm levels resulted in reduced movement, vomiting, and dyspnea. Exposure to 2000 ppm and higher methanol resulted in reduced blood pH during exposure but the blood pH recovered to normal during the recovery period. Pathological changes in the nervous system were noted at all test levels during exposure. Slight partial atrophy of the optic nerve was noted at 3000 ppm and higher but according to the author it was questionable if it was treatment related. No treatment related effects on vision, body weight, clinical signs, hematology, histopathology, and clinical chemistry were reported. One death was reported at day 5 in a monkey exposed at 5000 ppm. The NOAEL was <1000 ppm and the LOAEL was 1000 ppm based on pathological changes in the nervous system.

In a chronic inhalation study, monkeys were exposed to 0, 10, 100, or 1000 ppm methanol for 22 hours per day for a maximum of 2.5 years (NEDO, 1986, 1987). Eight monkeys were exposed at each level. Two monkeys at each treatment level (one in control) were sacrificed after 7 months of exposure. Three monkeys at each treatment level (two in control) were sacrificed after 19 months of exposure. The remainder three per group were sacrificed at 2.5 years. Parameters evaluated effects on vision, body weight, clinical signs, hematology, histopathology clinical chemistry, and survival. In the animals sacrificed after 7 months of exposure, no clinical signs were noted, but fatty degeneration of the liver at animal exposed at 100 or 1000 ppm methanol were observed. Slight pathological changes in the nervous system (hyperplasia of reactive astroglias in the white matter) were noted at the 7 month sacrifice in all treated animals. In the 1000 ppm monkey, slightly abnormal ECG and decreased blood pH were noted after 7 months of exposure. Histopathological examination after 19 months showed slight nervous system and liver effects at all treatment levels. Some nervous system effects (hyperplasia of reactive astroglias in the white matter) of questionable significance (reversible during recovery?) were also seen at the final sacrifice in all treatment groups (2.5 years). The nervous system effects were judged to be transient and reversible after exposure was terminated. Slight changes in liver and kidney were noted in monkeys exposed to 1000 ppm and kidney effects were reported at 100 ppm in the final sacrifice. ECG indicated slight myocardial effects and degeneration of the thalamus at 100 ppm and higher.

No changes in feed consumption, water consumption, body temperature, hematology, the cerebral cortex, and the peripheral nerves or in the retina, were noted. The NOAEL was <10 ppm and the LOAEL was 10 ppm based on the reported slight pathological changes in the nervous system.

The doses in this study were the basic for the doses used in the NEDO chronic and carcinogenicity studies in mice and rats.

Andrews et al. (1987) reported on a subchronic inhalation study of methanol in rats and monkeys. The targeted exposure concentrations were 0, 500, 2000, or 5000 ppm for 5 days per week, 6 hours per day for 4 weeks. Five male and five females Sprague Dawley rats and three male and three females *M. cynomolgus* monkeys were exposed at each test level. The methanol used was 99.85% pure. Body weights were measured weekly, and clinical observations were made twice daily. In addition, ophthalmic examinations were made pretest and at termination of the study. At termination organ weights were taken (adrenal gland, heart kidney, liver, lung spleen, testes, and ovaries). Approximately 35 tissues were examined microscopically.

The chamber mean measured concentrations were 550, 1980, and 5010 ppm. In rats, lacrimation and nasal discharge were observed in the treated animals, but only mucoid discharges were dose related. No treatment related clinical signs were noted in the monkey. No treatment related effects were noted in organ weight or body weights in any treated group of rats or monkeys. No ocular abnormalities were noted at the termination of the study. Gross and histological examination at termination of the study revealed no treatment related effects. No ocular abnormalities were noted in any of the high dose monkeys or rats when compared to controls.

Unfortunately, no blood measurements were made in this study. Blood methanol, and formic acid levels would have been useful. Other than signs of nasal irritation in rats exposed at 5000 ppm, the results of this study do not suggest that the top dose of 5000 ppm was toxic to either rats or monkeys.

It is interesting in the differences in response see in the subacute NEDO studies (NEDO, 1986, 1987) and the Andrews study (Andrews et al., 1987). The species of monkey used were the same in both studies. At the top dose (5000 ppm methanol) in the Andrews study (Andrews et al., 1987) no treatment-related effects were noted, while in the NEDO study (NEDO, 1986, 1987), in a range finding study the NOAEL was 3000 ppm methanol and the LOAEL (clinical signs, blood methanol levels, and blood formic acid levels) was 5000 ppm. One major difference is the length of exposure in these two studies. In the Andrew study (1987) exposure was 6 hours per day for 5 days per week for 4 weeks. In the NEDO study (NEDO, 1986, 1987), exposure was nearly continuous (21 hour daily for 14 days at 500 ppm and 21 days at 300 ppm). In the Andrews study there was a much greater likelihood that methanol did not build up over time, which happened in the NEDO study (see blood methanol and formic acid levels).

4.5.2 Rats

In the NEDO rat chronic inhalation study, exposure was for 20 hours per day for 12 months with concentrations of methanol of 0, 10, 100, or 1000 ppm. Fischer 344 rats were used, 20 per sex group. This study was run at the same time as an 24-month carcinogenicity in rats. The 24-month study is discussed in more detail in Chapter 8 on cancer. OECD testing guidelines were used, but no indication of GLP was mentioned. Parameters evaluated included clinical tests data (urine, hematological, and biochemical endpoints), body weight, feed consumption, survival and organ weights in addition to macroscopic and histological evaluations of the standard tissues.

No treatment-related effects were observed in clinical parameters, body weight, feed consumption, survival, histopathological, urinalysis, biochemical tests, and organ weights at 100 ppm and lower. Slight changes that may or may not have been treatment related were noted (body weight gain, liver, and spleen weights changes in rats exposed to 1000 ppm methanol). One rat died during the study and one was scarified (level not indicated). The NOEAL appeared to be 100 ppm with small body decrease rate of weight gain and possible organ weight treatment-related effects seen at the LOAEL of 1000 ppm.

Methanol effects testicular production of testosterone and testes morphology was also studied in folate-deficient rats exposed to 800 ppm methanol 6–20 hours per day, 5–7 days per week for up to 13 weeks. Exposure to 200 ppm methanol had no effect on testosterone production in either normal rats or folate-deficient rats. Methanol had no adverse effect on testes morphology at the end of 13 week exposure to 800 ppm methanol in either normal rats or folate-deficient rats. An increase incidence of testicular degeneration was noted in 18-month-old folate-deficient rats exposed to 800 ppm methanol for 13 weeks (20 hours per day 7 days per week) suggesting an age-related testes degeneration in these folate-deficient rats (Lee et al., 1991).

In a limited study, Sprague Dawley rats were exposed to 200, 2000, or 10,000 ppm for 6 hours per day for as long as 6 weeks. The rats were sacrificed 16 hours after last exposure and the lungs were removed and lavaged. The lavage fluid was analyzed for lactic dehydrogenase. No significant treatment-related effects were noted. No histopathological evaluation of the lung was conducted, but gross evaluation of the lung showed no indication of inflammation or irritation (White et al., 1983).

Folate-deficient rats were used successfully as a model to study methanol effects on the optic system in rats following inhalation exposure up to 3000 ppm methanol or orally at 3500 mg/kg. No changes in blood formic acid were seen in normal rats exposed to methanol, but blood formic acid was greatly increased at the same level of exposure in folate-deficient rats. Eight of 11 folate-deficient rats died after exposure to 3000 ppm methanol, but none of the normal rats exposed at the same concentration died. The blood formic acid increase in folate-deficient rats is similar to what is seen in humans following methanol exposure (Lee et al., 1994). The role of folate and difference in metabolism are discussed in Chapter 7 on metabolism.

In two similar inhalation studies, Sprague Dawley rats were exposed to methanol or a combination of methanol/toluene, toluene or methanol/gasoline, gasoline, 5 days per week 6 hours per day for 4 weeks (Poon et al., 1994, 1995). In the first study, methanol exposure was 300 or 3000 ppm. Mild to moderate changes in the thyroid were noted in all treated groups (methanol, toluene, and methanol/ toluene). Changes in the nasal cavity were also noted in high dose methanol group (Poon et al., 1994). In the second study, only the methanol level was changed (2500 ppm). The treatment-related effects were mild changes in the nasal cavity, hepatmegaly, and liver enzyme changes attributed primary to the gasoline in the gasoline only or combination of methanol/gasoline (Poon et al., 1995).

4.5.3 Mice

The design of the mouse chronic inhalation study conducted by NEDO was very similar to the chronic rat study just discussed (NEDO, 1986, 1987). Exposure was by inhalation for 20 hours per day for 12 months with concentrations of methanol of 0, 10, 100, or 1000 ppm. CRJ: B6C3F mice were used, 30 males and 30 females per group. This study was run at the same time as an 18-month carcinogenicity in mice, which is discussed in more detail in Chapter 8 on cancer. OECD testing guidelines were used, and again no indication that the study was conducted under GLP was mentioned. Ten mice per group were sacrificed at 6 months and the remaining 20 per group were scarified at 12 months. Clinical tests, plus checking for infection, body weight, feed consumption, and organ weights were evaluated. In addition, macroscopic and histological evaluations of the standard tissues were conducted. No treatment-related effects were observed in clinical tests, organ weights, or in histopathological examinations at 100 ppm or lower. Food consumption varied based on dose, time in the study, and body weight change. However, no clear treatment-related trend was observed. Slight changes that may be treatment related (weight gain, organ weight, and fatty liver degeneration) were observed in male mice exposed at 1000 ppm methanol. One mouse died during the study and one was sacrificed (100 ppm). The NOEAL appeared to be 100 ppm with small possible treatment-related increased body weight and fatty liver seen at the LOAEL of 1000 ppm.

If one compares the results of the three NEDO chronic inhalation studies conducted in monkeys, rats, and mice, the monkeys show effects at lower doses than the rat or mouse. The LOAEL is 10 ppm in monkeys, because of slight pathological changes in the nervous system and 1000 ppm in rats and mice because of slight changes in body and organ weights that may be treatment related.

4.5.4 Dogs

In an early inhalation study, four dogs were exposed to 450–500 ppm methanol for 8 hours per day 7days per week for 1 year. Hematogical, clinical signs, and ophthalmic exams were conducted. No adverse effects were reported (Sayers et al., 1942).

In another early limited inhalation study, two dogs were exposed at about 10,000 ppm methanol for about 3 minutes at hourly interval (8 time daily) for 100 days. Both dogs survived and show no clinical sign of treatment. Ophthalmic findings were negative (Sayers et al., 1944).

4.6 REPEAT EXPOSURE—ORAL

4.6.1 Rats

USEPA conducted a 90-day subchronic oral study in Sprague Dawley rats. Thirty rats per sex per dose were gavaged with 0, 100, 500, or 2500 mg/kg bw of methanol per day. Ten per sex per dose were sacrificed after 6 weeks. The remaining rats were scarified at the end of the study. Weekly body weights, food consumption, clinical signs of toxicity, opthalmological evaluation, and mortality were evaluated through out the study. Evaluation of blood, urine chemistry, gross, and microscopic tissue were conducted at sacrifice.

Brain weights were significantly less than the control at the high dose (2500 mg/kg). A slight but not statistically significant elevation of liver weight and liver enzymes (SGPT, SAP) were also noted in both male and female rats exposed at 2500 mg/kg. However, no treatment-related histopathological lesions were seen in the liver in rats exposed at 2500 mg/kg methanol. No treatment-related effects were seen in body weights, food consumption, and clinical signs of toxicity, opthalmological evaluation, mortality, as well as gross and microscopic tissue evaluation. The NOAEL was judged to be 500 mg/kg (USEPA, 1986).

In limited earlier studies, only biochemical changes in the brain were reported in five rats given 790 mg/kg orally for 7 days (Saad, 1976).

In another study, no significant changes in physiological parameters were noted in rats given 1% methanol in the drinking water for 6 months (Lington and Bevan, 1994).

4.6.2 Non-Human Primates

In a study to develop information on the monkey as a model of humans, six rhesus monkeys were given 2000 mg/kg bw methanol and then followed with five doses of 500 mg/kg methanol within 1.5–6 days to prolong the state of intoxication. Acidosis, accumulation of formic acid, and optic effects (little or no response to light and pathological effects) were reported suggesting that the monkey was a good model to study methanol poisoning in humans (Martin-Amat et al., 1977).

4.6.3 Mice

In an oral life time study in mice (Apaja, 1980), three levels of methanol in the drinking water were used as "special" control groups. All groups had 30 mice per sex. This study was designed to assess carcinogenicity of malonaldehyde in drinking water, but each of the three test levels of malonaldehyde released a different amount of methanol, which was used initially to stabilize malonaldehyde The three "special" methanol control groups had levels of 0.22%, 0.444%, or 0.898% methanol only in their drinking water. Therefore, this study could give information on the potential effects of repeat oral exposure of methanol in mice. No toxic effects due to methanol treatment were reported in female mice. This study is discussed in more detail in Chapter 8 on carcinogenicity.

4.7 REPEAT EXPOSURE—DERMAL

4.7.1 Mice

There are two dermal studies (Forbes et al., 1979; Apaja, 1980) that use methanol as a solvent/control. These studies are designed to assess carcinogenicity of other chemicals, but also could potentially give information on the effects of repeat dermal exposure of methanol in the solvent/controls. No toxic effects due to methanol treatment were reported. These studies are discussed in more detail in Chapter 8 on carcinogenicity.

4.8 AQUATIC TOXICITY

The acute toxicity to fish is listed in Table 4.4. Both static and flow through testing were used in more than a dozen fish species. In general, the results were similar with low acute toxicity being the general pattern of toxicity in fish.

The acute toxicity to aquatic invertebrate is in Table 4.5. In general, the results were similar (low acute toxicity), in the dozen or more species tested.

TABLE 4.4 Acute Aquatic Methanol Toxicity to Fish

Species	Exposure Time Hours	Concentration, mg/l	Comments	References
Bleak	96	LC_{50} = 28,000	Static	Bengtsson et al. (1984)
Goldfish	48	LC_{50} = 1,700		AQUIRE (1995)
Carp	48	LC_{50} = 28,000		AQUIRE (1995)
Carp	96	LC_{50} = 36,000		Dobson et al. (1993)
Armed bullhead	96	LC_{50} = 7,900–26,070	Static	Portmann and Wilson (1971)
Bluegill	24	LC_{50} = 19,230		Call et al. (1983)
	48	LC_{50} = 19,230		
	72	LC_{50} = 17.720		
	96	LC_{50} = 15,400		
Bluegill	24	LC_{50} = 19,100		Poirier et al. (1986)
	48	LC_{50} = 19,100		
	96	LC_{50} = 15,400		

TABLE 4.4 ***(Continued)***

Species	Exposure Time Hours	Concentration, mg/l	Comments	References
Bluegill	24	$LC_{50} = 19{,}230$		USEPA (1983)
	96	$LC_{50} = 15{,}500$		
Rainbow trout	24	$LC_{50} = 20{,}300$		Call et al. (1983)
	48	$LC_{50} = 20{,}100$		
	96	$LC_{50} = 19{,}000$		
Rainbow trout	24	$LC_{50} = 20{,}300$		Poirier et al. (1986)
	48	$LC_{50} = 20{,}100$		
	96	$LC_{50} = 20{,}100$		
Rainbow trout	96	$LC_{50} = 19{,}000$	Static	Mayer et al. (1986)
Rainbow trout	96	$LC_{50} = 21{,}000$	Fingerling	USEPA (1983)
Medaka	48	$LC_{50} = 41{,}000$		AQUIRE (1995)
	48	$LC_{50} = >10{,}000$	at 10°C	
	48	$LC_{50} = 1{,}400$	at 30°C	
Guppy	7 days	$LC_{50} = 10{,}860$	Static	Konemann (1981)
Fathead minnow	96	$LC_{50} = 28{,}200$		Vieth et al. (1983)
Fathead minnow	24	$LC_{50} = 28{,}400$		Call et al. (1983)
	48	$LC_{50} = 28{,}400$		
	72	$LC_{50} = 28{,}400$		
	96	$LC_{50} = 28{,}400$		
Fathead minnow	24	$LC_{50} = 29{,}700$		Poirier et al. (1986)
	48	$LC_{50} = 29{,}700$		
	96	$LC_{50} = 24{,}700$		
Fathead minnow	96	$LC_{50} = 29{,}400$		Brooke et al. (1986)
Sea bream	24	$LC_{50} = 13{,}000$		NEDO (1986)
	96	$LC_{50} = 12{,}000$		
Killifish	24	$LC_{50} = 35{,}000$		NEDO (1986)
	96	$LC_{50} = 28{,}000$		
Ear shell	24	$LC_{50} = 30{,}000$		NEDO (1986)
	96	$LC_{50} = 15{,}000$		
Zebra fish	96	$LC_{50} = 290–22{,}300$		Zeljenkova et al. (2006)
Golden orfe	48	$LC_{50} = 7900–10{,}000$	Static	Juhnke and Ludemann (1978)

For alga EC50 values (the concentration that reduces the maximum observed biodegradation rate by 50%) ranged from 540 mg/l for blue green alga to 28,400 mg/l for green alga (WHO, 1997).

There does not appear to be any chronic toxicity data in fish or aquatic invertebrate in the literature. Neutral compounds such as methanol have a low octanol–water partition and have a low acute toxicity. Low levels of methanol (0.1 ml/l) is used in some test protocol (40 CFR 797.1560, ASTM E1241-92) as a solvent.

TABLE 4.5 Acute Toxicity of Methanol to Invertebrates

Species	Exposure Time Hours	Concentration, mg/l	References
Water flea	48	NOEL = 10,000	Verschuren (1983)
Water flea	24	EC_{50}> => 10,000 (immobilization)	Kuhn et al. (1989)
	48	EC_{50} => 10,000 (immobilization)	
Water flea	24	EC_{50} = 23,500 (immobilization)	Rossini and Ronco (1996)
	48	EC_{50} = 22,200 (immobilization)	
Water flea	24	EC_{50} = 24,500	Bringmann and Kuhn (1982)
Water flea	24	EC_{50} = 21,402	Calleja et al. (1994)
Water flea	48	LC_{50} = 24,500	Randall and Knopp (1980)
Water flea	96	LC_{50} => 100	Ewell et al. (1986)
Water flea	3	LC_{50} = 38,000, LC_{50} = 41,000	AQUIRE (1995)
	18	LC_{50} = 19,500	
	24	LC_{50} => 10,000	
Water flea	48	LC_{50} = 13,240	Vaishnav and Korthals (1990)
Brine shrimp	24	LC_{50} => 10,000	Price et al. (1974)
Brine shrimp	24	LC_{50} = 10,000	Verschuren (1983)
Brine shrimp	24	LC_{50} = 1,578	Barahona-Gomariz et al. (1994)
24-h old	24	LC_{50} = 1,101	
48-h old	24	LC_{50} = 901	
72-h old			
Brine shrimp	24	LC_{50} = 43,574	Calleja et al. (1994)
Grass shrimp	18	LC_{50} = 21,900	AQUIRE (1995)

TABLE 4.5 (*Continued*)

Species	Exposure Time Hours	Concentration, mg/l	References
Tiger shrimp	24	$LC_{50} = 25{,}000$	NEDO (1986)
	96	$LC_{50} = 13{,}000$	
Brown shrimp	48	$LC_{50} = 1{,}975$	Portmann and Wilson (1971)
	96	$LC_{50} = 1{,}340$	
Glass shrimp	18	$LC_{50} = 21{,}900$	Bowman et al. (1981)
Flatworm	96	$LC_{50} = > 100$	Ewell et al. (1986)
Scud	96	$LC_{50} = > 100$	Ewell et al. (1986)
Scud	18	$LC_{50} = 19{,}350$	AQUIRE (1995)
Ramshorn snail	96	$LC_{50} = > 100$	Ewell et al. (1986)
Oligochaete	96	$LC_{50} = > 100$	Ewell et al. (1986)
Harpacticoid copepod	96	$LC_{50} = 12{,}000$	AQUIRE (1995)
Sowbug	96	$LC_{50} = > 100$	Ewell et al. (1986)
Mosquito	18	$LC_{50} = 20{,}000$	AQUIRE (1995)
Mussel	96	$LC_{50} = 15{,}900$	Helmstetter et al. (1996)
Cockle	48	$LC_{50} = 7{,}900$	Portmann and Wilson (1971)
	96	$LC_{50} = 7{,}900–2610$	
Rotifer	24	$LC_{50} = 35{,}884$	Calleja et al. (1994)

4.9 CONCLUSION

In this chapter, the acute and general repeat toxicity in animals has been reviewed as well as the aquatic toxicity. Methanol is generally classified as practically nontoxic or very low toxicity following acute exposure by the oral, dermal, and inhalation route in rodents. In general, the aquatic results were similar, with low acute toxicity being the general pattern of toxicity in fish and in aquatic invertebrate.

The lethal dose in humans is several folds lower than the lethal dose in animals making it obvious that the human is more sensitive to the acute effects of methanol than animals. The difference in the acute response in animals and humans is due to difference in the metabolism of methanol between animals and humans. The repeat methanol exposure data in rodents shows general toxicological response such as effects on body weight gain, and slight liver effects as treatment related effects. More information about chronic exposure of rodents to methanol is found in Chapter 8 on cancer. Developmental effects in rodents are found in Chapter 5. Interesting, effects were seen in repeat exposure rodent studies; above the acute lethal dose in humans (~1000 mg/kg) suggesting that repeat exposure data in rodents have limited relevance to humans.

The chapters in this book address the key methanol issues of metabolism in animals, reproductive and developmental results in rodents, and the potential issue of cancer in rodents. The other chapters help the understanding of the key issues in animal toxicity of methanol and how it relates to humans.

REFERENCES

Alder, P., Busche, W., and Gordononoff, T. (1938) Experimentelle untersuchungen uber dietoxizitat des methylalkohols. *Arch. Int. Pharmacol.* **59**, 416–430.

American Methanol Intstitute (2001) High production Volume Program on Methanol submitted to EPA 66p.

Andrews, L.S., Clary, J.J., Terrill, J.B., and Bolte, H.F. (1987) Subchronic inhalation toxicity of methanol. *J. Toxicol. Environ. Health* **20**, 117–124.

Apaja, M. (1980) Evaluation of toxicity and carcinogenicity of malonaldehyde. Acta Univ. Oul D55. *Anat. Pathol. Microbiol.* **8**, 1–59.

AQUIRE (1995) ECOTOX Database prepared for US EPA by Mid-Continent Ecology Division Duluth, MN.

Baer, G. (1898) Beitrag zur Kenntniss vergiftung mit verschiedenen alkoholen. *Arch. F. u Anat. Physiol, f. Physiol.* 283–296.

Barahona-Gomariz, M.V., Sanz-Barrera, F., and Sanchez-Foryun, S. (1994) Acute toxicity of organic solvents on Artemia salina. *Bull. Environ. Contam. Toxicol.* **52** (5), 766–771.

BASF AG. (1961) unpublished data cited by American Methanol Institute (2001) in High production Volume Program submission to USEPA 66p.

BASF AG. (1975) unpublished data cited by American Methanol Institute (2001) in High production Volume Program submission to USEPA 66p.

BASF AG. (1979) unpublished data cited by American Methanol Institute (2001) in High production Volume Program submission to USEPA 66p.

BASF AG. (1980) unpublished data cited by American Methanol Institute (2001) in High production Volume Program submission to USEPA 66p.

Bengtsson, B.E., Renberg, L., and Tarkpea, M. (1984) Molecular structure and auatic toxicity: An example with 1-carbon to 13-carbon aliphatic alcohols. *Chemosphere* **13** (5–6), 613–622.

Bowman, M.C., Oiller, W.L., Cairns, T., Gosnell, A.B., and Oliver, K.H. (1981) Stressed bioassay systems for rapid screening of pesticide residues Part 1 Evaluation of bioassay systems. *Arch. Environ. Contam. Toxicol.* **10**, 9–24.

Bringmann C. and Kuhn, R. (1982) Comparison of the toxicity thresholds of water pollutants to bacteria, algae and protozoa in the cell multiplication inhibition test. *Water Res.* **14**, 231–241.

Brooke, L.T., Call, D.J., Geiger, and Northcott, C.E. (Eds.) (1986) Acute toxicities of organic chemicals to fathead minnow (*Pimephales promelas*) Superior WI: Center for Lake Superior Environmental Studies, University of Wisconsin Superior, p19.

Call, D.J., Brooks, L.T., Ahamd, N., and Richter, J.E. (1983) Toxicity and metabolism studies with USEPA priority pollutants and related chemical. USEPA 600/3-83-095 PB83-263665.

Calleja MC., Persoone G., and Geladi P. (1994) Comparative acute toxicity of the first 50 milticentre evaluation of an in vivo cytotoxicity chemicals to aquatic non-vertebrates. *Arch. Environ. Contam. Toxicol.* **26**, 69–78.

Cameron, A.M., Nilsen, O.G., Haug, E., and Eik-Nes, KB. (1984) Circulating concentrations of testosterone, lutenizing hormone and follicle stimulating hormone rats after inhalation of methanol. *Arch. Toxicol.* **7**, 441–443.

Cameron, A.M., Zahlsen, K., Haug, E., Nilsen, O.G., and Eik-Nes, K.B. (1985) Circulating steroids in male rats following inhalation of n-alcohols. *Arch Toxicol.* **8** (Suppl), 422–424.

Carnegie-Mellon. (1981) unpublished data cited by American Methanol Institute (2001) in High production Volume Program submission to USEPA 66p.

Carpenter, C.P. and Smyth, H.F. (1946) Chemical burns of the rabbit cornea. *Am. J. Opthalmol.* **29**, 1363–1372.

Castro, G.D., Costantini, M.H., Delgado de Lavri, A.M., and Castro, J.A. (2002). Rat liver microsomal and nuclear activation of methanol to hydroxymethyl free radicals, *Toxicol. Lett.* **129** (3), 227–236.

Chikin, M.S., and Shadurskil. (1969) cited by American Methanol Institute (2001) in High production Volume Program submission to USEPA 66p.

Clay, K.L., Murphy, R.C., and Watkins, W.D. (1975) Experimental methanol toxicity in primates: Analysis of metabolic acidosis. *Toxicol. Appl. Pharmacol.* **34**, 49–61.

Cooper, J.R., and Felig, P. (1961) The biochemistry of methanol poisoning II. Metabolic acidosis in the monkey. *Toxicol. Appl. Pharmacol.* **3**, 202–209.

Cooper, R.L., Mole, M.L., Rehnberg, J.M., Goldman, J.M., McElroy, W.K., Hein, J., and Stoker, T.E. (1992) Effect of inhaled methanol on pituitary and testicular hormones in chamber acclimated non-acclimated rats. *Toxicology* **71**, 69–81.

Cornish, H.H. and Adefuin, J. (1967) Potentiation of carbon terachloride toxicity by aliphatic alcohols. *Arch. Environ. Health.* **14**, 447–449.

Deichmann, W.B. and Mergard, E.G. (1948) Comparative evaluation of methods employed to express the degree of toxicity of a compound. *J. Ind. Hyg. Toxicol.* **30**, 373–378.

Dobrzyska, I., Skrzydlewska, E., and Farbiszewski, R. (1999) Influence of trolox dervative and N-acetylcystiene on surface charge density of erythrocytes in methanol intoxicated rats. *Environ. Toxicol. Pharmacol.* **8**, 15–21.

Dobson, S.D., Malcolm, H.M., and Nielsen, I.R. (1993) Environmental Hazard Assessment: Methanol. Toxic Substance Division, Department of the Environment London, Vol TSD/17 45pp.

Dorman, D.C., Dye, J.A., Nassise, M.P., Ekuta, J., Bolon, B., and Medinsky, M.A. (1993) Acute methanol toxicity in minipigs. *Fundam. Appl. Toxicol.* **20**, 341–347.

DuPont (unknown) unpublished data cited by American Methanol Institute (2001) in High production Volume Program submission to USEPA 66p.

Eulner, H.H. and Gedicke, K.H. (1955) Uber die hautresorption von methyl alcohol. *Sammlung Vergiftungstalle.* **15**, 409–415.

Ewell, W.S., Gorusuch, J.W., Kringle, R.O., Robillard, K.A., and Spiegel, R.C. (1986) Simultaneous evaluation of the acute effects of chemicals on seven aquatic species. *Environ. Toxicol. Chem.* **5** (9), 831–840.

Forbes, P.D., Urbach, F., and Davies, R.E. (1979) Enhancement of experimental photocarcinogenesis by topical retinoic acid. *Cancer Lett.* **7**, 85–90.

Gilger, A.P. and Potts, A.M. (1955) Studies on the visual toxicity of methanol. V: The role of acidosis in experimental methanol poisoning. *Am. J. Ophthalmol.* **39**, 63–86.

Gilger, A.P., Potts, A.M., and Johnson, L.V. (1952) Studies on the visual toxicity of methanol II: The effect of parentally administered substance on the systemic toxicity of methyl alcohol. *Am. J. Ophthalmol.* **35** (Suppl), 113–126.

Goldenthal, E.I. (1971) A compilation of LD50 values in newborn and adult animals. *Toxicol. Appl. Pharmacol.* **18**, 153–207.

Haskell, C.C., Hilleman, S.P., and Gardner, W.R. (1921) The significance of the acidosis of methyl alcohol poisoning. *Arch. Int. Med.* **27**, 71–82.

Helmstetter, A., Gamerdinger, A.P., and Pruell, R.J. (1996) Acute toxicity of methanol to Mytilus edulis. *Bull. Environ. Contam. Toxicol.* **57** (4), 675–681.

Horton, V.L., Higuchi, M.A., and Richkert, D.E. (1992) Physiologically based pharmacokinetic model for methanol in rats, monkeys and humans. *Toxicol. Appl. Pharmacol.* **117**, 26–36.

Jacobs, G.A. (1990) Acute Toxicity Data. *J. Am. Coll. Toxicol.* **1**, 56.

Juhnke, V.I. and Ludemann, D. (1978) Results of testing of 200 chemicals compounds for acute toxicity with the Golden Orfe test. *WasserAbwasser Forch.* **11** (5), 161–164. (in German).

Kadilska, M.B. and Mason, R.P. (2000) Acute methanol intoxication generates free radicalsin rats: an SSR spin trapping investigation. *Free Radic. Biol. Med.* **28** (7), 106–114.

Kavet, R. and Knauss, K.M. (1990) The toxicology of inhaled methanol vapors. *Crit. Rev. Toxicol.* **21** (1), 22–50.

Kimura, E.T., Ebert, D.M., and Dodge, P.W. (1971) Acute toxicity and limits of solvent residue for sixteen organic solvents. *Toxicol. Appl. Pharmacol.* **19**, 699–704.

Konemann, H. (1981) Quantitative structure activity relationships in fish toxicity studies. *Toxicolgy.* **19**, 209–221.

Kuhn, R., Pattard, M., Pernak, K., and Winter, A. (1989) Results of the harmful effects of selected water pollutants (anilines, phenols, aliphatic compounds) to *Daphnia magna*. *Water Res.* **23**, 495–499.

Lee, E., Brady, A.N., Brabec, M.J., and Fabel, T. (1991) Effects of methanol vapors on testosterone production and testis morphology in rats. *Toxicol. Ind. Health.* **7**, 261–275.

Lee, E.W., Garner, C.D., and Terzo, T.S. (1994) Animal model for the study of methanol toxicity: Comparison of folate-reduced rat responses with published monkey data. *J. Toxicol. Environ. Health.* **41**, 71–82.

Lehmann, K.B. and Flury, F. (1943) *Toxicology and Industrial Solvents*, Williams and Wilkens, Baltimore.

Leo, A. (1927) Ueber chronische methylalkolvergiftung. *Biochem. Z.* **191**, 423–438.

Lington, A.W. and Bevan, C. (1994) Alcohols. In G.D. Clayton and F.E. Clayton (Eds.), *Patty's Industrial Hygiene and Toxicology*, Vol. 2 D in John Wiley & Sons, New York, pp. 2597–2609.

MAK. (1991) Occupational toxicants, critical data evaluation for MAK values and classification of carcinogen. In H Greim (Ed.), Vol. **16**, John Wiley-Vch, Weinheim, pp. 1–175.

Marc-Aurele, J. and Scheiner, G.E. (1960) The dialysis of ethanol and methanol: A proposed method for the treatment of massive intoxication by ethyl or methyl alcohol. *J. Clin. Invest.* **19**, 802–807.

Martin-Amat, G., Tephly, T.R., McMartin, K.E., Nakar, A.B., Hayreh, M.S., Hayreh, S.S., Baumbach, G., and Cancilla, P. (1977) Methyl alcohol poisoning: II. Development of a model for ocular toxicity in methyl alcohol poisoning using the rhesus monkey. *Arch. Opthalmol.* **95**, 1847–1850.

Mashbitz, N.L., Sklianskaya, Hayreh, S.S., Hayreh, M.S., and Tephly, T.R. (1936) Relative toxicity of acetone, methyl alcohol and their mixtures. Their action on white mice. *Ind. Hyg. Toxicol.* **8**, 117–122.

Mayer, F.L., Mayer, K.S., and Ellersieck, M.R. (1986) Relationship and survival to other endpoints in chronic toxicity tests with fish. *Environ. Toxicol. Chem.* **5** (8), 737–748.

McCord, C.P. (1931) Toxicity of methyl alcohol (methanol) and inhalation, following skin absorption. *Ind. Eng. Chem.* 931–936.

Munch, J.C. (1972) Aliphatic alcohol and alkyl esters: narcotic and lethal potencies to tadpoles and rabbits. *Ind. Med. Surg.* **412**, 31–33.

Munch, J.C. and Schwartze, E.W. (1925) Narcotic and toxic potency of aliphatic alcohols upon rabbits. *J. Lab. Clin. Med.* **10**, 985–996.

New Energy Development Organization (NEDO) (1986) Toxicological research of methanol as a fuel for power station. Tokyo, Japan. September.

New Energy Development Organization (NEDO) (1987) Toxicological research of methanol as a fuel for power station. Tokyo, Japan. October.

Parthasarathy, N.J., Kumar, R.S., Manikandan, S., and Devi, R.S. (2006) Methanol-induced oxidative stress in rat lymphoid organs. *J. Occup. Health.* **48**, 20–27.

Poirier, S.H., Knuth, M.L., Anderson-Buchou, C.D., Lima, A.R., and Shubat, P.J. (1986) Comparative toxicity of methanol and dimethylformamide to fresh water fish. *Bull. Environ. Contam. Toxicol.* **37** (4), 815–621.

Poon, R., Chu, I.H., Bjarnason, S., Polvin, M., Vincent, R., Miller, R.B., and Vallli, V.E. (1994) Inhalation toxicity study of methanol, toluene and methanol toluene mixtures in rats, effects of 28 day exposure. *Toxicol. Ind. Health.* **10**, 231–245.

Poon, R., Chu, I.H., Bjarnason, S., Vincent, R., Polvin, M., Miller, R.B., and Vallli, V.E. (1995) Short term inhalation toxicity study of methanol, gasoline and methanol gasoline in rats. *Toxicol. Ind. Health.* **11**, 343–361.

Portmann, J.E. and Wilson, R.W. (1971) Then toxicity of 140 substances to brown shrimp and other marine animals London Ministry of Agriculture Fisheries, and Foods (Shellfish Information leaflet No 331).

Potts, A.M. (1955) The visual toxicity of methanol: VI the clinical aspect of experimental methanol poisoning treated with a base. *Amer. J. Ophthamol.* **39**, 86–92.

Potts, A.M, Praglin, J., Farkas, I.S., Lowell, M.S., Orbison, L., and Chickering, D. (1955) Studies on the visual toxicity of methanol. VIII. Additional observation on methanol poisoning in the primate test object. *Am. J. Ophthamol.* **40**, 76–83.

Price, K.S., Waggy, G.T., and Conway, R.A. (1974) Brine shrimp bioassay and seawater BOD of petrochemicals. *J. Water Pollut. Control Fed.* **46** (1), 63–77.

Randall, T.L. and Knopp, P.V. (1980) Detoxification os specific organic substances by wet oxidation. *J. Water Pollut. Control Fed.* **52** (8), 2117–2150.

Registry of Toxic Effects of Chemical Substances (RTECS) (1994) DHHS (NIOSH).

Roe, O. (1982) Species differences in methanol poisoning. *CRC Crit. Rev. Toxicol.* **10**, 276–287.

Rossini, G.D.B. and Ronco, A.E. (1996) Acute toxicity bioassay using Daphnia obtusa as a test organism. *Environ. Toxicol. Water Qual.* **11** (3), 255–268.

RTECS (Registry of Toxic Effects of Chemical Substances). (1994). DHHS (NIOSH).

Saad, S.F. (1976) Effects of aliphatic alcohols on gamma-aminobutyric acid levels in the cerebral hemispheres of rats. *Soc. Occup. Med.* **4**, 499.

Sayers, R.R., WP, Yant, W.P., Schrenk, H.H., Chornyak, J., Pearce, S.J., Patty, F.A., and Linn, J.G. (1942) Methanol poisoning. I. Exposure of dogs to 450–500 ppm methanol vapors in air. Report of Investigations RI 3617 U. S. Dept. of Interior, Bureau of Mines. pp. 1–10.

Sayers, R.R., Yant, W.P., Schrenk, H.H., Chornyak, J., Pearce, S.J., Patty, F.A., and Linn, J.G. (1944) Methanol poisoning. II. Exposure of dogs for brief periods eight times daily to high concentrations of methanol vapor in air. *J. Ind. Hyg. Toxicol.* **26**, 255–259.

Scott, J.B., Smith, F.A., and Garman, R.H. (1979) Exposure to mice to CH_2CL_2 and CH_3OH and in combination. *Toxicol. Appl. Pharmacol.* **48**, 105A.

Skrzydlewska, E. and Farbiszewski, R. (1996) Disminished antioxidant defense potential of liver, erythrocytes and serum, from rats with subacute methanol intoxication. *Vet. Human. Toxicol.* **38**, 429–433.

Skrzydlewska, E. and Farbiszewski, R. (1998) Lipid peroxidation and antioxidant status in the liver, erythrocytes and serum of rats after methanol intoxication. *J. Toxicol. Envirn. Health.* **53**, 637–649.

Skrzydlewska, E., Witek, A., and Farbiszewski, R. (1998) The comparison of the antioxidant defense potential of brain to liver of rats after methanol ingestion. *Comp. Biochem. Physiology.* **120**, 289–294.

Smith, E.N. and Taylor, R.T. (1982) Acute toxicity of methanol in the folate deficient acatasemic mouse. *Toxicology.* **25**, 271–287.

Smyth, H.F. (2001) unpublished Carnegie-Mellon data cited by American Methanol Institute in High Production Volume Program submission to USEPA 66p.

Smyth, H.F., Seaton, J., and Fischer, L. (1941) The single dose toxicity of some glycol and derivatives. *J. Ind. Hyg. Toxicol.* **23**, 253–258.

Union Carbide (1994) Unpublished data cited by American Methanol Institute (2001) in High production Volume Program submission to USEPA 66 p.

USEPA (1983) *Toxicity and Metabolism Studies with USEPA Priority Pollutants and Related Chemicals in Freshwater Organisms*, US Environmental Protection Agency, Washington DC.

USEPA (1986) Rat oral subchronic toxicity study with methanol. Office of Solid Waste Washington DC.

USEPA (2008) Acute Toxicity Categories for pesticide products. CFR 40 Part 156. 62 Subpart B (7-1-08 Edition).

Vaishnav, D.D. and Korthals, E.T. (1990) Comparative toxicities of selected industrial chemicals to microorganisms and other aquatic organisms. *Arch. Environ. Contam. Toxicol.* **19** (4), 624–628.

Verschuren, K. (1983) *Handbook of Environmental Data on Organic Chemicals*, Van Nostrand Reinhold Co, New York.

Veith, G.D., Call, D.J., and Brooke, L.T. (1983) Structure-toxicity relationship for fathead minnow, *Pimephales promelas*, narcotic industrial chemicals. *Can. J. Fish Aquat. Sci.* **40** (6), 743–748.

Von Burg, R.J. (1994) Toxicology update. *Appl. Toxicol.* **14** (4), 5: 9–13.

Weil, C. (1972) Statistics vs safety factors and scientific judgment in the evaluation of safety for man. *Toxicol. Appl. Pharmacol.* **21**, 454–463.

Welch, H. and Slocum, G.G. (1943) cited by American Methanol Institute (2001) in High production Volume Program submission to EPA 66p.

Wesse, H. (1928) cited by Treon JF. Alcohols. Fassett DW, Irish DD. Editors. Pattv's Industrial Hygiene and Toxicology. Revised Second Edition. Vol. 2 John Wiley and Sons, New York. (1963) pp. 1409–1422.

White, L.R., Marthinsen, A.B.L., Richards, R.J., Eik-Nes, K.B., and Nusen, O. G. (1983) Biochemical and cytological of rat lung after inhalation of methanol vapor. *Toxicol. Lett.* **17**, 1–5.

Witte, R. (1931) cited by Treon JF. Alcohols. Fassett DW, Irish DD. (Eds.) Pattv's Industrial Hygiene and Toxicology. Revised 2nd edition. Vol. 2 John Wiley & Sons, New York. (1963) pp. 1409–1422.

World Health Organization (WHO) (1997) Methanol Environmental Health Criteria 196 180p.

Youssef, A.F., Weiss, B., and Cox, C. (1993) Neurobehavioral toxicity of methanol reflected by operant running. *Neurotoxicol. Teratol.* **15**, 223–227.

Zeljenkova, D., Koviznych, J.A., and Szabova, E. (2006) Acute toxicity of selected chemicals in adult zebrafish (*Danio Rerio*) and it's early life stages: The comparative study. *Reprod. Toxicol.* **22** (2), 285.

5 Developmental and Reproductive Toxicology of Methanol

JOHN M. ROGERS,[1] JEFFREY S. GIFT,[2] and STANLEY BARONE, JR[3]

[1]Toxicity Assessment Division, National Health and Environmental Effects Research Laboratory, Office of Research and Development, United States Environmental Protection Agency, Research Triangle Park, NC, USA

[2]Hazardous Pollutant Assessment Group, National Center for Environmental Assessment, Office of Research and Development, United States Environmental Protection Agency, Research Triangle Park, NC, USA

[3]National Center for Environmental Assessment, United States Environmental Protection Agency, Research Triangle Park, NC, USA

5.1 INTRODUCTION

Methanol is a well-known human toxicant, producing ocular toxicity and death following acute exposure by a number of routes (Tephly and Mcmartin, 1984; Kavet and Nauss, 1990), but human developmental toxicity has not been reported. Methanol has been identified as a rodent teratogen in studies undertaken because of the potential for occupational exposures or increased use of this alcohol in vehicle fuels (Nelson et al., 1985; Rogers et al., 1993; Bolon et al., 1993). Subsequent developmental and reproductive studies in monkeys were more equivocal in their findings, albeit at lower exposure levels (Burbacher et al., 1999a, b, 2004a, b). This chapter reviews the literature describing the developmental and reproductive toxicity of methanol and germane to understanding the pathogenesis and underlying mechanisms of these

The Toxicology of Methanol, First Edition. Edited by John J. Clary.

effects in laboratory animals. The metabolism of methanol is described elsewhere in this volume and must be considered in evaluating putative mechanisms through which this alcohol might produce dysmorphogenesis in offspring following maternal exposure. *In vitro* approaches, principally whole embryo culture, that have been important in elucidating the mechanisms of developmental toxicity of methanol and metabolites will also be considered here.

5.2 REPRODUCTIVE TOXICITY

There are no published studies on the effects of methanol exposure on male or female reproductive function in humans. There have been a few studies on the effects of methanol on male reproductive parameters in rats, which are discussed in this section. There have been many more studies on the effects of methanol on prenatal and early postnatal development in experimental animals, and these studies are discussed in later sections.

In a study of the testicular effects of methanol, Cameron et al. (1984) exposed male Sprague-Dawley rats to methanol vapor, 8 hours/day, 5 days/week for 1, 2, 4, and 6 weeks at 0, 200, 2000, or 10,000 ppm. The authors examined the possible effects of methanol on testicular function by measuring blood levels of testosterone, luteinizing hormone (LH), and follicular stimulating hormone (FSH) using radioimmunoassay. The most significant changes were in blood testosterone levels of animals exposed to 200 ppm methanol, the lowest concentration evaluated. At this exposure level, animals exposed for 6 weeks had testosterone levels that were 32% of those seen in controls. The lack of a dose–response was not necessarily an indication that the observed effect was not related to methanol, because higher concentrations of methanol could be causing other effects (e.g., liver toxicity), which can have an impact on testosterone levels. Male rats exposed to 10,000 ppm methanol for 6 weeks displayed blood levels of LH that were about three times higher than those exposed to air.

A follow-up study report by the same research group (Cameron et al., 1985) described the exposure of male Sprague-Dawley rats, 6 hours/day for either 1 day or 1 week, to methanol, ethanol, *n*-propanol, or *n*-butanol. Animals were sacrificed immediately after exposure or after an 18-hour recovery period and testosterone, LH, and corticosterone were measured in serum. The data were consistent with the ability of these aliphatic alcohols to cause a transient reduction in the formation of testosterone. Except in the case of *n*-butanol, rapid recovery from these deficits can be inferred from the 18-hour postexposure data.

Lee et al. (1991) exposed 8-week-old male Sprague-Dawley rats to 0 or 200 ppm methanol, 8 hours/day, 5 days/week, for 1, 2, 4, or 6 weeks to assess effects on testosterone production. There was no effect on serum testosterone, gross structure of reproductive organs, or weight of testes and seminal vesicles. Lee et al. (1991) also studied the *in vitro* effect of methanol on testosterone production from isolated testes, but saw no effect on testosterone formation either with or without stimulation with human chorionic gonadotropin. In another experiment from the same report, testicular histopathology was evaluated to determine if methanol exposure produced lesions indicative of changing testosterone levels; the effects of age and folate status were also assessed. Groups of 4-week-old male Long-Evans rats were given diets containing either adequate or reduced folate levels. Some rats were exposed to 0, 50, 200, or 800 ppm methanol starting at 7 months of age while others were exposed to 0 or 800 ppm methanol at 15 months of age. Methanol exposures were conducted continuously for 20 hours/day for 13 weeks. The authors reported that visual toxicity and acidosis developed in rats that were fed the low-folate diet and exposed to methanol. No methanol-related testicular lesions or changes in testes or body weight were evident at 10 months of age in rats that were fed either the folate sufficient or deficient diet. Likewise, no methanol-induced lesions were observed in 18-month-old rats that were fed diets with adequate folate. The incidence but not severity of age-related testicular lesions was increased in the 18-month-old rats fed with folate-deficient diets. Observations of subcapsular vacuoles in germinal epithelium were increased in the 800 ppm group; one rat in the 800 ppm group had atrophied seminiferous tubules and another had Leydig cell hyperplasia. The role of folate in the metabolism and toxicity of methanol is discussed in greater detail later in this chapter.

Cooper et al. (1992) conducted experiments on the effects of an acute (6-hour) methanol exposure on serum testosterone and pituitary hormone levels in male rats. In addition, these investigators examined the potential role of animal husbandry on the results observed. Male Harlan Sprague-Dawley rats at 90 days of age were either acclimated to the inhalation chambers and handled for 2 weeks prior to beginning exposures or were started on the exposures with no acclimation. Test animals received a single 6-hour inhalation exposure to 0 (sham air-exposed), 200, 5000, or 10,000 ppm methanol. One group of rats was killed immediately following exposure and the remaining animals were killed 18 hours later (24 hours after the start of exposure). In the control groups, serum hormones were generally higher in the nonacclimated versus the acclimated rats, suggesting an effect of handling on hormone concentrations. Unlike the results of Cameron et al. (1984), Cooper et al. (1992) did not observe a change in serum testosterone levels in either acclimated or nonacclimated rats at 200 ppm methanol. Immediately following exposure, nonacclimated rats exposed to 5000 ppm methanol exhibited an apparent ~40% reduction in serum LH compared to nonacclimated controls, but acclimated controls had serum LH levels similar to those measured in acclimated rats exposed to 5000 ppm methanol. At 24 h, serum testosterone was lower in acclimated rats exposed to 10,000 ppm methanol but higher in nonacclimated rats exposed to 5000 ppm. In the second experiment, Cooper et al. (1992) measured serum methanol, testosterone, LH, and prolactin in 90-day-old male rats immediately after exposure to 5000 ppm methanol for 1, 3, or 6 hours. These studies illustrate that, while 1, 3, or 6 hours methanol exposures can impact rat serum hormone levels, the magnitude and direction of changes may depend on the handling of the animals.

5.3 DEVELOPMENTAL TOXICITY

There are no published studies that report developmental toxicity in humans following methanol exposure. Because of its proposed use as an alternative automobile fuel in the late 1980s, a number of studies were undertaken in rats, mice, and non-human primates to elucidate the potential for methanol to cause developmental toxicity if exposures occurred during pregnancy and/or the early postnatal period. In this section, we present the findings of those studies, which include both structural and behavioral effects of maternal methanol exposure during pregnancy and in the early postnatal period. Subsequent sections examine the pathogenesis of observed developmental disorders as well as the role of methanol metabolism and maternal folate status.

5.3.1 Rats

The effects of inhaled methanol during pregnancy in the rat were first studied by Nelson et al. (1985). Sprague-Dawley rats were exposed to 5000, 10,000, or 20,000 ppm methanol for 7 h/day. The two lower concentrations were administered daily from GD 1 to GD 19, while the 20,000 ppm exposures were administered daily from GD 6 to GD 15. Blood methanol levels were assessed in nonpregnant females under the same exposure conditions. Signs of maternal toxicity were slight and were observed only at the highest concentration. Peak blood methanol levels were approximately 1.3, 2.0, and 8.7 mg/ml after exposure to 5000, 10,000, or 20,000 ppm, respectively. Maternal exposure to 20,000 ppm methanol resulted in decreased fetal weight and significant increases in external, visceral, and skeletal malformations. Skeletal malformations were the most prevalent and included abnormalities of the basicranium and the vertebra, including an increase in the incidence of fetuses with a rib on the seventh cervical vertebra. There was also a low incidence of exencephaly, encephalocele, hydrocephalus, and various anomalies of the cardiovascular and urinary systems at the highest exposure level. At 10,000 ppm, there were slight, statistically insignificant increases in these same anomalies, as well as a significant decrease in fetal weight, so this was determined to be the Lowest Observed Adverse Effect Level (LOAEL). No maternal or developmental parameters were affected at 5000 ppm methanol, the No Observed Adverse Effect Level (NOAEL) for this study.

The New Energy Development Organization of Japan (NEDO) (1987) sponsored a teratology study in Sprague-Dawley rats that included an evaluation of postnatal effects in addition to standard prenatal endpoints.

A total of 36 pregnant females/group were exposed to 0, 200, 1000, or 5000 ppm (0, 262, 1310, and 6552 mg/m^3) methanol vapors (reagent grade) on GD 7–17 for 22.7 hours/day. Statistical significance of results was evaluated by *t*-test, Mann-Whitney U test, Fisher's exact test, and/or Armitage's χ^2 test.

In the prenatal portion of the NEDO (1987) study, reduced body weight gain and food and water intake during the first 7 days of exposure were reported for dams in the 5000 ppm group. On GD 20, 19–24 dams/group were killed for evaluation of litters and individual fetuses. Adverse reproductive and fetal effects were limited to the 5000 ppm group and included an increase in late-term resorptions, decreased live fetuses, reduced fetal weight, and increased frequency of litters with fetal malformations, variations, and delayed ossifications. Malformations or variations included defects in ventricular septum, thymus, vertebrae, and ribs.

Postnatal effects of methanol inhalation were evaluated in the remaining 12 dams/group that were permitted to deliver and nurse their litters. Effects were only observed in the 5000 ppm group, and included a 1-day prolongation of the gestation period and reduced post-implantation survival, number of live pups/litter, and survival on postnatal day (PND) 4. When the delay in parturition was considered, methanol treatment had no effect on attainment of developmental milestones such as eyelid opening, auricle development, incisor eruption, testes descent, or vaginal opening. There were no adverse body weight effects in offspring from methanol-treated groups. The weights of some organs (brain, thyroid, thymus, and testes) were reduced in 8-week-old offspring exposed to 5000 ppm methanol during prenatal development.

NEDO (1987) contains an account of a two-generation reproductive study that evaluated the effects of pre- and postnatal methanol (reagent grade) exposure (20 hours/day) on reproductive and other organ systems of Sprague-Dawley rats. The F_0 generation (30 males and 30 females per exposure group) was exposed to 0, 10, 100, or 1000 ppm from 8 weeks old to the end of mating (males) or to the end of lactation period (females). The F_1 generation was exposed to the same concentrations from birth to the end of mating (males) or to weaning of F_2 pups 21 days after delivery (females). Males and females of the F_2 generation were exposed from birth to 21 days old (one animal/sex/litter was exposed to

8 weeks of age). NEDO (1987) noted reduced brain, pituitary, and thymus weights, and early testicular descent in the offspring of F_0 and F_1 rats exposed to 1000 ppm methanol. The early testicular descent is believed to be an indication of earlier fetal development as indicated by the fact that it was correlated with increased pup body weight. However, no histopathologic effects of methanol were observed. As discussed in the report, NEDO (1987) sought to confirm the possible compound-related effect of methanol on the brain by carrying out an additional study in which Sprague-Dawley rats were exposed to 0, 500, 1000, or 2000 ppm methanol from the first day of gestation through the F_1 generation. Brain weights were measured in 10–14 offspring/sex/group at 3, 6, and 8 weeks of age. As illustrated in Table 5.1, brain weights were significantly reduced in 3-week-old males and females exposed to $\geq$1000 ppm. At 6 and 8 weeks of age, brain weights were significantly reduced in males exposed to $\geq$1000 ppm and females exposed to 2000 ppm.

The postnatal effects of methanol exposure during late gestation in Long-Evans rats were studied by Infurna and Weiss (1986). Methanol was administered at 2% in the drinking water for 3 days from GD 15 to GD 17 or GD 17 to GD 19. Daily methanol intake averaged about 2.5 g/kg body weight. On PND 1, suckling behavior was tested by placing the pups with an anesthetized dam and recording the latency to finding the nipple and beginning to suckle. On PND 10, the ability of the pups to locate nesting material from their home cage was tested. Methanol exposure did not affect maternal weight gain or fluid intake; nor did it affect litter size, birth weight, postnatal survival or growth, or day of eye opening. Offspring of methanol-exposed dams required longer than controls to begin suckling on PND 1 (latency of approximately 90 s vs 60 s in controls). Ability to locate material from the home nest in a plexiglass grid was also affected in methanol-exposed pups on PND 10. The methanol-exposed pups required about twice as long as controls to find their nest material, and their initial direction was more often incorrect than that of controls.

Stanton et al. (1995) exposed pregnant rats to 15,000 ppm methanol via inhalation for 7 hour/day on GD 7–19. Daily peak maternal blood methanol concentrations declined from 3.8 mg/ml after the first

TABLE 5.1 Brain Weights of Rats Exposed to Methanol Vapors During Gestation and/or Lactation (NEDO, 1987)

		Brain Weight, g					
Offspring Age	Sex	0 ppm	200 ppm	500 ppm	1000 ppm	2000 ppm	5000 ppm
3 weeks	Male	1.45 ± 0.06	ND	1.46 ± 0.08	1.39 ± 0.05^a	1.27 ± 0.06^b	ND
	Female	1.41 ± 0.06		1.41 ± 0.07	1.33 ± 0.07^c	1.26 ± 0.09^b	
6 weeks	Male	1.78 ± 0.07	ND	1.74 ± 0.09	1.69 ± 0.06^c	1.52 ± 0.07^b	ND
	Female	1.68 ± 0.08		1.71 ± 0.08	1.62 ± 0.07	1.55 ± 0.05^b	
8 weeks[d]	Male	1.99 ± 0.06	ND	1.98 ± 0.09	1.88 ± 0.08^c	1.74 ± 0.05^b	ND
	Female	1.85 ± 0.05		1.83 ± 0.07	1.80 ± 0.08	1.67 ± 0.06^b	
8 weeks[e]	Male	2.00 ± 0.05	2.01 ± 0.08	ND	1.99 ± 0.07	ND	1.81 ± 0.16^c
	Female	1.86 ± 0.08	1.91 ± 0.06		1.90 ± 0.08		1.76 ± 1.09

[a] $p < 0.05$.

[b] $p < 0.001$, as calculated by the authors. Values are means ± S.D.

[c] $p < 0.01$.

[d] Exposed throughout gestation and F_1 generation.

[e] Exposed on gestational days 7–17 only.

exposure to 3.1 mg/ml after the final exposure. Offspring body weight was reduced by about 5% on PND 1, but the behavioral tests used, including motor activity (PND 13–21, 30, 60), olfactory learning (PND 18), behavioral thermoregulation (PND 20–21), T-maze learning (PND 23–24), acoustic startle response (PND 24, 60), reflex modification audiometry (PND 60), passive avoidance (PND 72), and visual-evoked potentials (PND 160) did not detect any difference between methanol-exposed and control groups.

The effects of methanol exposure during early pregnancy in the rat were studied by Cummings (1993). Rats were dosed orally with up to 3.2 g/kg/day methanol on GD 1–8 and groups of animals were killed on GD 9, 11, or 20. The decidual cell response (DCR) technique, in which the lining of the uterus is surgically stimulated to induce proliferation, was also applied. Reductions in pregnant uterus weight and implantation site weights were observed on GD 9 and effects on the DCR suggested that methanol impeded uterine decidualization. No other effects on viability or development of conceptuses were noted.

5.3.2 Mice

Rogers et al. (1993) studied the effects of methanol exposure (1000, 2000, 5000, 7500, 10,000, 15,000 ppm) during pregnancy in CD-1 mice under conditions similar to those used for rats by Nelson et al. (1985). No maternal toxicity was attributed to methanol although the exposure procedure *per se* reduced maternal weight gain in all groups, including the filtered air-exposed mice, compared to unhandled controls. An additional unexposed control group was food-deprived for 7 hour/day similar to the food deprivation experienced by the exposed groups. Food-deprivation accounted for some, but not all, of the effect of exposure on maternal weight. CD-1 mice were found to be more sensitive to the developmental toxicity of methanol than were Sprague-Dawley rats. Cleft palate, exencephaly, and skeletal defects were observed at 5000 ppm and above. The incidence of fetuses with ribs on the seventh cervical vertebra was increased in a dose-related fashion at 2000 ppm and above. The skeletal anomalies observed, including cervical vertebral anomalies and cervical ribs, were similar to those observed in rats by Nelson et al. (1985). There was a significant increase in resorptions at 7500 ppm and above and an increase in fully resorbed litters at 10,000 and 15,000 ppm. Despite malformations at lower exposure levels, fetal weight was reduced only at 10,000 and 15,000 ppm. Interestingly, the induction of cleft palate and exencephaly by methanol were found to be nonindependent events in this study. The incidence of cleft palate among fetuses without exencephaly was 21.9%, while the incidence of cleft palate among fetuses with exencephaly was only 11.1%. Yasuda et al. (1991), studying TCDD, reported a similar decrease in susceptibility to cleft palate among fetuses with exencephaly.

The effects of methanol administered by oral gavage during pregnancy in CD-1 mice were also studied by Rogers et al. (1993a). Mice were given twice daily dosages of 2 g/kg methanol, 7 hours apart, on GD 6–15. Effects observed were similar to those observed following inhalation exposure (cleft palate, exencephaly, skeletal defects, and resorptions), and quantitatively the peak blood level and incidence of effects were similar to those observed after an inhalation exposure to 10,000 ppm methanol for 7 hours/day.

The developmental phase specificity for the adverse effects of exposure to inhaled methanol in pregnant CD-1 mice has been examined. Bolon et al. (1993) exposed pregnant mice to 10,000 ppm methanol (7 hour/day) for 10 days on GD 6–15; to 5000, 10,000, or 15,000 ppm for 3 days on GD 7–9; or to 10,000 or 15,000 ppm for 3 days on GD 9–11. Gestation day 17 fetuses were examined for external and visceral anomalies, but skeletal exams were not done. Exposure to 10,000 or 15,000 ppm on GD 6–15 resulted in cleft palate, exencephaly, hydronephrosis, tail and digit defects, and increased resorptions. Exposure to 10,000 or 15,000 ppm on GD 7–9 resulted in exencephaly, cleft palate, hydronephrosis, and ocular and tail defects. Exposure to these same concentrations on GD 9–11 resulted in cleft palate and digit and tail defects. Exposure to 15,000 ppm methanol on GD 7–8 also caused exencephaly. Single or two-day (7 hours/day) inhalation exposures of pregnant CD-1 mice to 10,000 ppm methanol during the period of GD 5–13 were carried out by Rogers and Mole (1997). Two-day exposures were carried out beginning on each of GD

6–12 (e.g., the latest exposure was on GD 12–13) and single-day exposures were on each of GD 5–9. Peak blood methanol concentration after a single exposure was about 4 mg/ml, and blood levels returned to baseline within about 24 hours. Two-day exposures on GD 6–7 or 7–8, and single-day exposure on GD 7 resulted in some fully resorbed litters. With single-day exposure, the number of resorptions per litter was the highest on GD 7. The period of susceptibility to methanol-induced cleft palate was broad, with single exposures on any of gestation days 5–9 or two-day exposures on GD 6–7 through 11–12 eliciting this effect. Peak sensitivity to cleft palate occurred with two-day exposure on GD 7–8 or one-day exposure on GD 7. The basis for this rather early critical period (induction as early as GD 5) for methanol-induced cleft palate is unknown at present, but is unusual for an agent with a short biological half-life. Exencephaly occurred with two-day exposure on GD 6–7 through GD 8–9 (peak GD 6–7) and one-day exposure on GD 5 through GD 8 (peak GD 7). The peak and breadth of critical periods for skeletal defects were progressively later going from more anterior to more posterior structures. Thus, the critical periods were GD 5 for exoccipital defects, GD 5 or 6 for atlas defects and GD 7 for axis defects, lower cervical defects, and supernumerary (lumbar) ribs. Single-day exposure on GD 5 resulted in a significant increase in the incidence of fetuses with 25 presacral vertebrae (26 is normal), while single-day exposure on GD 7 resulted in an increased incidence of fetuses with 27 presacral vertebrae. The stage-specific patterns of developmental sensitivity to cleft palate, exencephaly, cervical rib, atlas (vertebra C1) defects, and axis (vertebra C2) defects following single-day exposures are illustrated in Figure 5.1.

The skeletal abnormalities observed by Rogers and Mole (1997), including splits and duplications of the atlas and axis, ribs on cervical vertebra seven and abnormal number of presacral vertebrae were suggestive of disruption of embryo segmentation and/or segmental identity. These skeletal malformations were examined in greater detail by Connelly and Rogers (1997). Methanol (5 g/kg) was administered orally to CD-1 mice on GD 7 and fetuses were collected on GD 18. Anatomical landmarks identifying specific cervical vertebrae were examined, including the tubercula anterior normally found on cervical

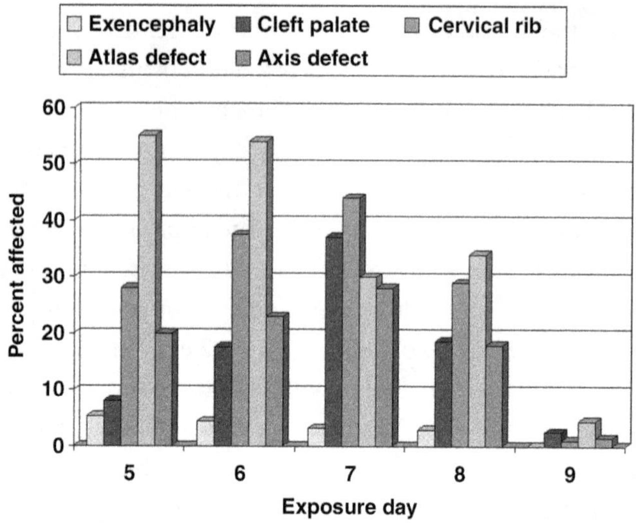

FIGURE 5.1 Incidence of prevalent malformations in fetuses following single exposures of pregnant CD-1 mice to 10,000 ppm methanol for 7 hours on the indicated day. Fetuses were examined on gestation day 17. *Source*: From Rogers and Mole (1997).

vertebra six (C6) and various foramina and other features evident in disarticulated vertebrae. The number of free (e.g., ventrally unattached) ribs and ribs attached to the sternum were counted, and ribs found on C7 were categorized as partial or full (i.e., attached to the sternum). Methanol caused homeotic shifts of segment identity. Specifically, a posteriorization of vertebral elements, especially in the cervical region, was observed (Figure 5.2). That is, certain of these vertebrae had structural features normally found on the next vertebra posteriad. Methanol-treated fetuses often had cartilaginous tubercula anterior on C5 rather than their normal position on C6, and full ribs attached to the sternum were observed on C7. Further, morphological abnormalities of the atlas and axis (fusions, splits, and duplications) gave the appearance of disrupted segmentation. The biological basis for these striking skeletal alterations is unknown, but similar phenotypes have been observed in mice in which homeobox gene function has been ablated.

A prominent mouse model of the fetal alcohol syndrome has been developed in the C57BL/6J mouse (Sulik et al., 1981; Sulik and Johnston 1983; Webster et al., 1983). To examine potential similarities

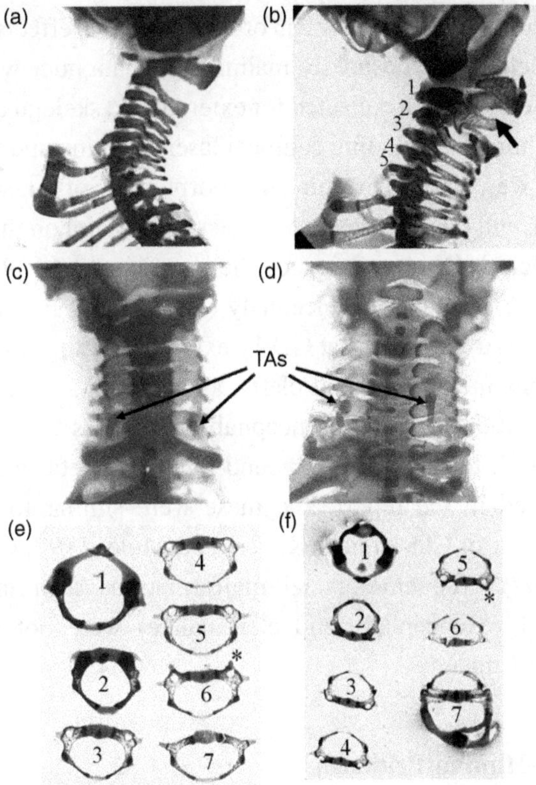

FIGURE 5.2 Alizarin red (bone) and Alcian blue (cartilage) stained cervical skeletal regions of fetuses from control (a, c, e) and methanol treated (b, d, f) CD-1 mice. Pregnant mice were treated with 4 or 5 g/kg by oral gavage on gestation day 7, and fetuses were examined on gestation day 17. Note splits and fusion of the atlas and axis (arrow in b) and unilateral displacement (to C5) of the tubercular anterior (TA) in the ventral view of the treated fetus in d (compare to the control in c, where both TAs are on C6). Also, compare disarticulated cervical vertebrae from a control fetus (e) to those from a methanol-treated fetus (f). TAs (asterisks) on C6 in the control are on C5 in the treated fetus, and C7 from the treated fetus has a full rib attached to it. *Source*: From Connelly and Rogers (1997).

in the effects of methanol and ethanol, Rogers et al. (2004) gave pregnant female C57BL/6J mice two intraperitoneal (i.p.) injections of methanol on GD 7, the most sensitive day for induction of craniofacial malformations by ethanol. The injections were given 4 hours apart to provide a total dosage of 0, 3.4, and 4.9 g/kg. Some pregnant mice were killed on GD 7, 8, or 9 in order to examine effects on embryos following methanol exposure. Remaining pregnant mice were killed on GD 17 and fetuses were collected for external and skeletal examination. Imaging techniques including confocal laser scanning and fluorescence microscopy were used to examine the morphology of fetuses excised at GD 7, GD 8, and GD 9. Apparent effects of methanol on the embryonic forebrain included a narrowing of the anterior neural plate, missing optic vesicles, and holoprosencephaly (Figure 5.3). Fetuses collected from methanol-treated dams at GD 17 exhibited a range of craniofacial malformations including facial clefts, anophthalmia, microphthalmia, varying severities of holoprosencephaly, and gross facial dysgenesis (Figure 5.4). A total of 15 compound-related skeletal malformations were tabulated in the report, and these were similar to the skeletal alterations seen in CD-1 mice (Rogers and Mole, 1997; Connelly and Rogers, 1997). The craniofacial malformations seen in methanol-treated CD-1 (exencephaly and cleft palate) were not observed in the C57BL/6J mice.

5.3.3 Non-Human Primates

Burbacher et al. (1999a, b) carried out toxicokinetic and reproductive/developmental studies of methanol in *Macaca fascicularis* monkeys that were published by the Health Effects Institute (HEI) in a two-part monograph. Some of the data were subsequently published in the open scientific literature (Burbacher et al., 2004a, b). Two cohorts of 12 monkeys/group were exposed to low levels (relative to the previously discussed rodent studies) of 0, 200, 600, or 1800 ppm methanol vapors for 2.5 hours/day, 7 days/week, during a premating and mating period (180 days combined) and throughout the entire gestation period (168 days). The study included an evaluation of maternal reproductive performance and tests to assess infant postnatal growth and newborn health, reflexes, behavior, and development of visual, sensorimotor, cognitive, and social behavioral function. Blood methanol levels, clearance, and the appearance of formate were also examined.

With regard to reproductive parameters, there was a statistically significant decrease in length of pregnancy in all treatment groups.

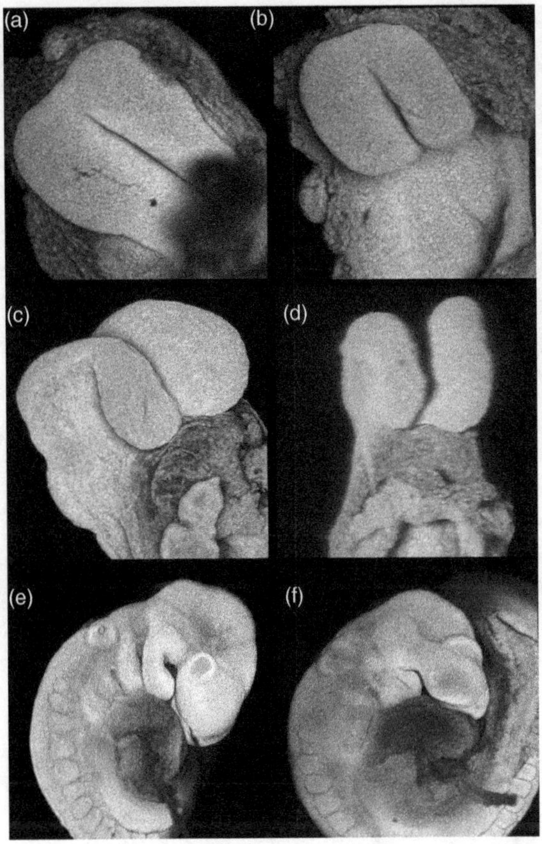

FIGURE 5.3 CD-1 mouse embryos from control (a, c, e) or methanol-treated C57BL/6J mouse dams (b, d, f) on gestation day 7 (a, b), 8 (c, d), and 9 (e, f). Embryos were fixed and surface stained with acridine orange and photographed on a fluorescence microscope. Pregnant mice were injected i.p. with 4.9 g/kg methanol, split in two doses 4 hours apart on gestation day 7. On gestation day 7, 2 hours after the second dose, the anterior neurectoderm of the treated embryos (b) was narrow in comparison to controls (a). At gestation day 8, further narrowing of the anterior neural folds and lack of the optic pit is evident (d) compared to the control (c). By gestation day 9, micro/anophthalmia is apparent in the methanol-treated embryo (f), along with poorly developed branchial arches and small forebrain, compared to the control (e).

Maternal menstrual cycles, conception rate, and live birth index were all unaffected. There were also no effects on maternal weight gain or clinical toxicity among the dams. The decrease in pregnancy length was largely due to complications of pregnancy requiring Cesarean section

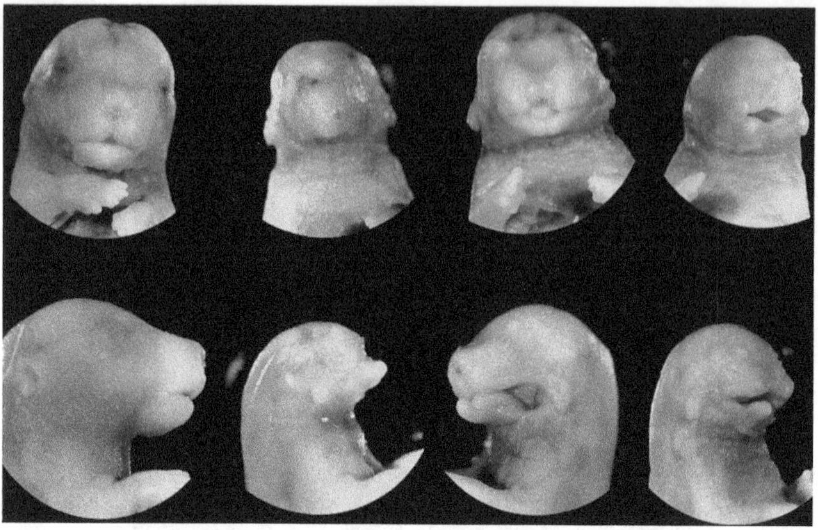

FIGURE 5.4 Examples of the variety of the severe craniofacial dysgenesis observed in fetuses from C57BL/6J dams treated i.p. with 4.9 g/kg methanol on gestation day 7. Controls are at left, top, and bottom. All other fetuses are from methanol-treated dams and exhibit holoprosencephaly, facial clefts, agnathia, and gross facial dysgenesis.

(C-section) deliveries in the methanol exposure groups, in response to signs of difficulty in the pregnancy.

While pregnancy duration was virtually the same in all exposure groups, there were some indications of increased pregnancy duress in methanol-exposed monkeys. C-sections were done in two monkeys from the 200-ppm group and two from the 600-ppm group due to vaginal bleeding. A monkey in the 1800 ppm group also received a C-section after experiencing nonproductive labor for three nights. In addition, signs of prematurity were observed in one infant from the 1800 ppm group that was born after a 150-day gestation period. Fetal parameters such as crown-rump length and head circumference were not different among the groups. Infant growth and tooth eruption were unaffected by prenatal methanol exposure.

In later life, two females out of the total of nine offspring in the 1800 ppm group exhibited a wasting syndrome at 12 and 17 months of age. Food intake was normal and no cause of the syndrome could be determined. Necropsies revealed gastroenteritis and severe malnourishment.

No infectious agent or other pathogenic factor could be identified. Thus, it appears that a highly significant toxicological effect on postnatal growth can be attributed to prenatal methanol exposure at 1800 ppm. The effects observed were in the absence of appreciable increases in maternal blood formate levels.

In summary, the Burbacher et al. (1999b, 2004a, b) studies suggest that methanol exposure may cause reproductive effects, manifested as a shortened mean gestational period due to pregnancy complications that precipitated delivery via a C-section, and developmental neurobehavioral effects which may be related to the shortened gestational period. The effect on gestation length at 200 ppm is questionable, as the decrease in gestational length was marginally significant and largely the result of human intervention (C-section). Also, this effect did not appear to be dose related, the greatest gestational period decrease having occurred at the lowest (200 ppm) exposure level.

Neurobehavioral function was assessed in 8–9 infants/group during the first 9 months of life (Burbacher et al., 1999b, 2004a). Although results in 7/9 tests were negative, two effects were possibly related to methanol exposure. The visually directed reaching (VDR) test is a measure of sensorimotor development and assessed the infants' ability to grasp for a brightly colored object containing an applesauce-covered nipple. Beginning at 2 weeks after birth, infants were tested five times/day, 4 days/week. Performance on this test, measured as age from birth at achievement of test criterion (successful object retrieval on 8/10 consecutive trials over two testing sessions), was reduced in all treated male infants. Statistical significance was obtained in the 1800 ppm group when males and females were evaluated together and in the 600 ppm group for males only. However, there were no significant differences between responses and/or variances among the dose levels for males and females combined, or for males only. Yet there was a significant dose–response trend for females only ($p = 0.0265$). The extent to which VDR delays were due to a direct effect of methanol on neurological development or were secondary to the methanol-induced decrease in gestation length is not clear. Studies of reaching behavior have shown that early motor development in preterm human infants without major developmental disorders differs from that of full-term infants (Fallang et al., 2003), and clinical studies have indicated that the quality of reaching and grasping behavior in preterm infants is generally less than that in full-term infants (Fallang et al., 2003; Plantinga et al., 1997).

At 190–210 days of age, the Fagan test of infant intelligence was conducted. This paradigm makes use of the infant's proclivity to direct more visual attention to novel stimuli rather than familiar stimuli. The test measures the time infants spend looking at familiar versus novel items. Deficits in the Fagan task can qualitatively predict deficits in intelligence quotient (IQ) measurements assessed in children at later ages (Fagan and Singer, 1983). Control monkey infants in the Burbacher et al. (1999b, 2004a) study spent more than 62% of their time looking at novel versus familiar monkey faces, while none of the treated monkeys displayed a preference for the novel faces. Results of this test were not different among control and exposed groups. A cohort effect appeared to reduce the statistical power of this analysis. Because of the longer latency between birth and assessment, these results would not be expected to be confounded by the postulated methanol-induced decrease in gestation length of the exposed groups. Negative results were obtained for the remaining seven tests that evaluated early reflexes, gross motor development, spatial and concept learning and memory, and social behavior.

5.3.4 Summary of Developmental Toxicity Findings in Experimental Animals Exposed to Methanol by Inhalation

While the oral gavage and drinking water routes of exposure were used in some of the studies described earlier, the most complete data sets in terms of dose–response relationships that allow one to posit NOAELs and LOAELs have used inhalation exposures. Findings from these studies are summarized in Table 5.2. The lowest NOAEL from among the tabulated inhalation studies is for effects of methanol on postnatal brain weight in rats from the NEDO (1987) study. The NOAEL for this effect was 500 ppm methanol. However, where studies were carried out in a similar fashion (e.g., in the rat by Nelson et al. (1985) and in the mouse by Rogers et al. (1993)), the mouse appears to be more sensitive

TABLE 5.2 Summary of Developmental Toxicity Studies of Methanol Toxicity in Experimental Animals by Inhalation Exposure

Species, Strain	Doses/Duration	NOAEL, ppm	LOAEL, ppm	Effects	Reference
Sprague-Dawley rat	0, 5 k, 10 k, 20 k ppm, 7 hours/day either GD 1–19 or GD 7–15	5,000	10,000	Lower fetal weight, visceral and skeletal defects, including cervical ribs	Nelson et al. (1985)
Sprague-Dawley rat	0, 200, 1 k, 5 k ppm, 22.7 hours/day, GD 7–17	1,000	5,000	Resorptions, fetal death, malformations, variations, and delayed ossifications.	NEDO (1987)
Sprague-Dawley rat (two-generation study)	0, 10, 100, or 1 k ppm, 20 hours/day; F_1 – birth to end of mating (M) or weaning (F); F_2 – birth to 8 weeks	100	1,000	Lower brain, pituitary, and thymus weights at 8, 16, and 24 weeks postnatal in F_1 and at 8 weeks in F_2	
Sprague-Dawley rat (follow-up of brain weights in F_1)	0, 500, 1 k, 2 k ppm; GD 0 through F_1 generation	500	1,000	Lower brain weight at 3 and 6 weeks (males). Lower brain and cerebrum weight at 8 weeks (males)	

(Continued)

TABLE 5.2 (*Continued*)

Species, Strain	Doses/Duration	NOAEL, ppm	LOAEL, ppm	Effects	Reference
CD-1 mouse	0, 1 k, 2 k, 5 k, 7.5 k, 10 k, 15 k ppm, 7 hours/day, GD 6–15	1,000	2,000	Resorptions, cervical ribs, cleft palate, exencephaly; lower fetal weight, delayed ossification	Rogers et al. (1993)
CD-1 mouse	0, 10 k ppm, 2 days during GD 6–13 or single days during GD 5–9	NA	10,000	Cleft palate, exencephaly, skeletal malformations	Rogers and Mole (1997)
Long-Evans rat	0 or 15 k ppm, 7 hours/day GD 7–19	NA	15,000	Lower birth weight	Stanton et al. (1995)
Long-Evans rat	0 or 4.5 k ppm, GD 10 to PND 21.	NA	4,500	Subtle cognitive deficits	Weiss et al. (1996)
Macaca fascicularis	0, 200, 600, or 1.8 k ppm, 2.5 hours/day, 7 d/week, premating, mating, and gestation	ND	ND[a]	Shortened gestation (no dose–response), neurobehavioral deficits (lower performance in VDR test) may be related to premature births	Burbacher et al. (1999a, b, 2004a, b)

ND, not determined due to study limitations such as small number of animals/time point/exposure level; NA, not applicable.

[a]Gestation resulted in a shorter period of gestation in dams exposed to as low as 200 ppm (263 mg/m^3). However, because of uncertainties associated with these results, including clinical intervention and the lack of a dose–response, a definitive NOAEL or LOAEL could not be determined from this study.

than the rat. The NOAEL for teratogenicity in the mouse was 1000 ppm based on the induction of cervical ribs in methanol exposed mouse fetuses (Rogers et al., 1993), but no evaluation of possible effects of maternal methanol exposure on brain growth in the mouse has been performed.

5.3.5 Pathogenesis of Methanol-Induced Birth Defects

5.3.5.1 Whole Animal Studies The pathogenesis of methanol-induced neural tube defects in CD-1 mouse embryos and fetuses was examined by Bolon et al. (1994). Following maternal exposure to 15,000 ppm methanol (6 hours/day) on GD 7–9, 15% of fetuses exhibited cephalic dysraphism on GD 17. The severity of the defects ranged from encephalocele to exencephaly with or without facial clefting, anencephaly, and holoprosencephaly. Further, microcephaly was observed among methanol-exposed fetuses without neural tube defects. Embryos examined on GD 8.5 following maternal methanol exposure (15,000 ppm) on GD 7–8 had swollen and poorly elevated cephalic neural folds compared to controls. Reductions in the mitotic indices of the neuroepithelium (55% reduction) and the underlying mesoderm (47% reduction) were reported. Additional dams were exposed to 15,000 ppm methanol on GD 7–9 and embryos examined on GD 9.5 and 10.5. The methanol-exposed embryos exhibited delayed rotation, microcephaly, and edema as well as anterior neural tube patency. Nile blue sulfate staining of embryos to visualize apoptosis on GD 8.5, 9.5, or 10.5 revealed no difference in the staining pattern between methanol-treated and control embryos.

To determine the effects of methanol on cell proliferation and histological changes in fetuses of treated dams, Degitz et al. (2004a) used an identical experimental protocol to that of Rogers et al. (2004), administering 0, 3.4, or 4.9 g methanol/kg in distilled water i.p. (split doses, 4 hours apart) to C57BL/6J mice on GD 7. Embryos were collected at various times on GD 8 and GD 10. Embryos from dams exposed to 4.9 g/kg methanol on GD 7 exhibited, on GD 8, reductions in anterior mesenchyme subjacent to the mesencephalon and the base of the prosencephalon (embryonic forebrain). The optic pits were often lacking; where present the optic cup ectoderm was thin. These finding are in agreement with the observations of methanol-induced anophthalmia and microphthalmia in this mouse strain by Rogers et al. (2004). Degitz et al. (2004a) also reported that there were fewer neural crest cells in the mid- and hindbrain regions in methanol-treated embryos on GD 8.

Degitz et al. (2004a) also examined the effects of methanol on development of the cranial nerves and ganglia using immunohistochemical visualization of these structures. Dose-related abnormalities in the development of the cranial nerves and ganglia were seen after maternal methanol treatment on GD 7. On GD 9, extensive cell death was evident in areas populated by the neural crest, including the forming cranial ganglia. Development of ganglia V, VIII, and IX was reduced in a dose-related manner. Ganglia VII and X exhibited reduced development only at the highest dose (4.9 g/kg). Also at this dose, the brain and face were poorly developed and the brachial arches were reduced in size or virtually absent. Flow cytometry of dissociated nuclei from the head regions of the embryos at GD 8 did not show an effect on the proportion of cells in S-phase.

5.3.5.2 In Vitro Studies Abbott et al. (1995) examined patterns of cell death in rat and mouse embryos exposed to methanol in culture. Early somite stage embryos were exposed to dysmorphogenic concentrations of methanol and observed for cell death using a modified Feulgen whole-mount staining procedure, which allows nuclei throughout the embryo to be viewed *in situ*. Confirming the results of Andrews et al. (1993a), methanol retarded growth and development (including delayed neural tube closure) of both rat and mouse embryos, and mouse embryos were affected at lower concentrations. Increased cell death was observed in specific regions of the forebrain, visceral arches, and optic placodes in both species.

The effects of methanol on palatogenesis *in vitro* were examined and compared to effects of ethanol by Abbott et al. (1994). Mid-craniofacial tissues from GD-12 CD-1 mouse embryos were cultured in serum-free medium for 4 days. Exposure to 0–20 mg/ml methanol for 6 hours, 12 hours, 1 or 4 days resulted in a concentration and time-dependent decrease in the incidence and completeness of palatal fusion. A specific effect on cell proliferation was evidenced by reductions in total DNA content at concentrations of methanol that did not reduce protein content, and by decreased incorporation of ^{3}H-thymidine in palatal mesenchyme. Ethanol appeared to be a more potent inhibitor of palatogenesis *in vitro* than methanol. This system probably does not, however, model the mechanisms inducing cleft palate *in vivo*, since the critical period of susceptibility *in vivo* is during an earlier developmental stage.

In a study designed to elucidate quantitative and qualitative strain differences in response to methanol, GD 8 C57BL/6J and CD-1 mouse embryos with 5–7 somites were cultured in the presence of increasing concentrations of methanol (Degitz et al., 2004b). Embryos were cultured in 0, 1, 2, 3, 4, or 6 mg methanol/ml for 24 hours and evaluated for morphological development. Cell death was increased in both strains in a developmental stage- and region-specific manner at 4 and 6 mg/ml after 8 hours of exposure. The proportions of cranial region cells in S-phase were significantly decreased following 8- and 18-hour exposures to 6 mg methanol/ml culture medium. After 24 hours of exposure, C57BL/6J embryos had significantly decreased total protein at 4 and 6 mg/ml. Significant developmental effects were seen at 3, 4, and 6 mg/kg, with eye dysmorphology being the most sensitive endpoint. CD-1 embryos had significantly decreased total protein at 3, 4, and 6 mg/kg, but developmental effects were seen only at 6 mg/kg. It was concluded that the C57BL/6J embryos were more severely affected by methanol in culture than were CD-1 embryos.

5.3.6 Folate Deficiency—A Susceptibility Factor for Methanol Developmental Toxicity?

Humans and other primates are susceptible to the effects of methanol exposure associated with formate accumulation because they have lower levels of hepatic tetrahydrofolate-dependent enzymes involved in formate oxidation. Pregnant women are often of marginal or deficient folate status. Tetrahydrofolate-dependent enzymes and critical pathways that depend on folate, such as purine and pyrimidine synthesis, may also play a role in the developmental toxicity of methanol. Studies of rats and mice fed folate-deficient diets have identified developmental defects and adverse effects on reproductive performance, implantation, and fetal growth. Inhibition of folate cellular transport has been associated with severe developmental abnormalities including neural tube defects, cleft-lip and cleft-palate, cardiac septal defects, and eye defects (Antony, 2007). Folate deficiency has been shown to exacerbate some aspects of the developmental toxicity of methanol in mice and rats, and these studies are discussed in this section.

Sakanashi et al. (1996) tested the influence of dietary folic acid intake on various reproductive and developmental effects observed in CD-1 mice exposed to methanol. Starting 5 weeks prior to breeding and continuing for the remainder of the study, female CD-1 mice were fed folic acid free diets supplemented with 400 (folic acid deficient, FAD), 600 (marginal), or 1200 (folic acid sufficient, FAS) nmol folic acid/kg diet. Females were bred after being fed their respective diets for 5 weeks. On GD 6–15, pregnant mice in each of the diet groups were gavaged twice daily with 2.0 or 2.5 g/kg/day methanol (total dosage of 4 or 5 g/kg/day). There was an approximately 50% reduction in maternal liver and plasma folate levels on GD 18 in mice fed FAD versus FAS diets in both the methanol exposed and unexposed groups. The oral methanol doses of 4–5 g/kg/day resulted in an increase in cleft palate in mice fed sufficient folic acid diets, as well as an increase in resorptions and a decrease in live fetuses per litter. They did not observe an increase in exencephaly in the FAS group at these doses. In animals fed the FAD diet and dosed at 5 g/kg methanol/day, there was a threefold increase in the percentage of litters affected by cleft palate (86.2% vs 34.5%) and a 10-fold increase in the percentage of litters affected by exencephaly (34.5% vs 3.4%) compared to mice fed the FAS diet and given the same dose of methanol. Plasma and liver folate levels at GD 18 within each dietary group were not significantly different between exposed versus unexposed mice. However, these measurements were taken 3 days after the last methanol dose. Dorman et al. (1995) observed a transient decrease in maternal red blood cell (RBC) and conceptus folate 2 hours after inhalation exposure to 15,000 ppm methanol on GD 8. Thus, it is possible that short-term reductions in available folate during GD 6–15 may have affected fetal development.

Fu et al. (1996) reported studies done in the same laboratory and used a similar study design and dosing regimen as Sakanashi et al. (1996), but exposed the pregnant mice to only the higher 2.5 g/kg/day BID methanol dose (total dosage of 5.0 g/kg/day) on GD 6–10. They measured maternal liver and plasma folate levels on GD 18 and observed similar, significant reductions in these levels for the FAD versus FAS mice. However, Fu et al. (1996) also measured fetal liver folate levels at GD 18. Significantly, the maternal FAD diet had a greater impact on fetal liver folate than maternal liver folate levels on GD 18. Another key finding in this study was that methanol exposure during GD 6–10 had similar fetotoxic effects, including cleft palate, exencephaly, and resorptions, as the same level of methanol exposure administered during GD 6–15 (Sakanashi et al., 1996; Rogers et al., 1993). This is consistent with the hypothesis (Rogers and Mole, 1997) that the critical period for methanol-induced cleft palate and exencephaly in CD-1 mice is within GD 5–9.

Aziz et al. (2002) investigated the potential role of folic acid in methanol-induced developmental neurotoxicity in rats. Female albino Wistar were fed FAD or FAS diets for 14–16 weeks, after which liver folate levels were estimated and females exhibiting a significantly low folic acid level were mated. Throughout their lactation period, dams of both the FAD and FAS groups were given 0%, 1%, 2%, or 4% v/v methanol via drinking water, equivalent to approximately 480, 960, and 1920 mg/kg/day. Pups were exposed to methanol via lactation from PND 1–21. Liver folate levels were determined at PND 21 and neurobehavioral parameters (motor performance using the spontaneous locomotor activity test and cognitive performance using the conditioned avoidance response (CAR) test), and neurochemical parameters (dopaminergic and cholinergic receptor binding and dopamine levels) were measured at PND 45. The expression of GAP-43, a protein primarily localized to growth cone membranes, was examined using immunohistochemistry and western blot analysis.

Lower body weight gain was observed at PND 7, 14, and 21 in animals exposed to 2% or 4% methanol in the FAD group but only at 4% methanol in the FAS group. Liver folate levels in the FAD group were decreased by 63% in rats prior to mating and 67% in pups on PND 21.

These authors hypothesized that methanol would cause a depletion in dopamine levels and degeneration of the dopaminergic nigrostriatal pathway. Consistent with this hypothesis, they found dopamine levels were significantly decreased (32% and 51%) in the striatum of rats in the FAD group treated with 2% and 4% methanol, respectively. In the FAS group, a significant decrease (32%) was observed in the 4% methanol-exposed group.

Methanol treatment at 2% and 4% was associated with significant increases in activity in a spontaneous locomotor activity test in the FAS group (13% and 39%, respectively) and more notably, in the FAD group (33% and 66%, respectively) when compared to their respective controls. The authors suggest that alterations in locomotor activity may be due to a significant alteration in dopamine receptors and disruption in neurotransmitter availability. Dopamine receptor (D_2) binding in the hippocampus of the FAD group was increased by 34% in the 1% methanol group but was decreased by 20% and 42% in the 2% and 4% methanol groups, respectively. In the FAS group, D_2 binding was significantly increased by 22% and 54% in the 2% and 4% methanol groups.

At PND 45, performance in the CAR test in FAD rats exposed to 2% and 4% methanol was decreased by 48% and 52%, respectively, compared to non exposed animals fed the same diet. In the FAS group, the CAR was only decreased in the 4% methanol-exposed animals, and only by 22% compared to their controls. Aziz et al. (2002) suggest that the impairment in CAR of the methanol-exposed FAD pups may be due to alterations in the number of cholinergic (muscarinic) receptor proteins in the hippocampal region of the brain. Muscarinic receptor binding was significantly increased in the 2% (20%) and 4% (42%) methanol-exposed group in FAD animals, while FAS group animals had a significant increase in cholinergic binding only in the 4% methanol exposed group (21%).

Immunohistochemistry showed an increase in the expression of GAP-43 protein in the dentate granular and pyramidal cells of the hippocampus in 2% and 4% methanol-exposed offspring of animals fed the FAD diet, while offspring of animals fed the FAS diet showed increased expression only in the 4% methanol-exposed group. Western blot analysis confirmed a higher expression of GAP-43 in the 2% and 4% methanol-exposed FAD rats.

The Aziz et al. (2002) study provides evidence that hepatic tetrahydrofolate is an important factor in methanol-induced developmental neurotoxicity in rodents. The effects of methanol (increased spontaneous locomotor activity, decreased CAR, disturbances in dopaminergic and cholinergic receptors, and increased expression of GAP-43 in the hippocampal region) were more pronounced in FAD as compared to FAS rats. Thus, alterations in available folic acid, particularly to the conceptus, could have significant impacts on the developing fetus apart from the influence it is presumed to have on formate removal. These results do not implicate any particular proximate teratogen, as folate deficiency can increase levels of methanol, formaldehyde and formate (Medinsky et al., 1997). The immature blood–brain barrier and an inefficient drug-metabolizing enzyme system make the developing brain a sensitive target organ for the effects of methanol.

5.3.7 Role of Methanol and Metabolites in the Developmental Toxicity of Methanol

Andrews et al. (1993a) examined the direct toxicity of methanol to CD-1 mouse and Sprague-Dawley rat embryos developing in culture. In rat embryos, methanol concentrations of 8 mg/ml culture medium and above resulted in decreased growth and development, and 12 mg/ml resulted in dysmorphogenesis in 66% of live embryos as well as 53% embryo mortality. In comparison, methanol concentrations of 4 mg/ml culture medium affected growth and development of mouse embryos, and dysmorphogenesis was observed in 58% of embryos cultured in the presence of 6 mg/ml. Importantly, in both the rat and the mouse, concentrations of methanol required to cause developmental toxicity *in vitro* were similar to peak maternal blood levels following developmentally toxic exposures *in vivo* (rats, Nelson et al., 1985; mice, Rogers et al., 1993). Brown-Woodman et al. (1995) reported similar findings for the effects of methanol on Sprague-Dawley rat embryos in culture, with concentrations of 286.5 μmol/ml (9.17 mg/ml) being developmentally toxic. These results demonstrated that methanol does not require maternal metabolism to elicit developmental toxicity and suggest that the greater sensitivity of the mouse to the developmental toxicity of methanol is at least in part due to an intrinsic difference in embryonal sensitivity. However, they do not address the question of whether the embryo itself might metabolize sufficient amounts of methanol to toxic metabolites.

Adult human acute toxicity of methanol has been ascribed to accumulation of the metabolite formate. The study of formate developmental toxicity in intact rodents is precluded because of rapid metabolic elimination, but it has been tested in rodent embryos in culture. Andrews et al. (1993b) cultured presomite rat embryos for 48 hours in serum containing concentrations of Na^+-formate ranging from 0 to 1.6 mg/ml. The starting pH of the culture medium was 8.13 and was not affected by the addition of Na^+-formate. Since metabolism of methanol to formate *in vivo* would produce a metabolic acidosis, the effects of lowering pH in addition to formate exposure were tested. At pH 8.13, embryo development and growth was affected by Na^+-formate only at the highest concentration, 1.6 mg/ml (23.5 mM formate). However, decreasing media pH by the addition of HCl exacerbated the effect of formate, such that at pH 6.5 most developmental parameters were affected at 0.8 mg Na^+-formate/ml culture medium (11.8 mM formate). Confirming the results of Andrews et al. (1993b), Brown-Woodman et al. (1995) reported that concentrations at or above 18.66 mM formic acid or Na^+-formate were developmentally toxic to rat embryos *in vitro*, but that Na^+-formate was less toxic due to the exacerbating effect of low pH in formic acid-treated cultures. The developmental toxicities of Na^+-formate and formic acid were subsequently compared in mouse and rat embryos (Andrews et al., 1995). Rat and mouse embryos exposed to either agent for 24 hours exhibited concentration-dependent reductions in growth and development and increased incidences of malformed embryos. In contrast to results of previous studies with methanol, no difference in species sensitivity to Na^+-formate or formic acid was observed. Various developmental parameters were adversely affected at concentrations of 11.8 mM and above for both forms of this metabolite. The concentrations of formate found to be toxic to rat or mouse embryos growing *in vitro* are relevant to human risk assessment, because blood formate concentrations in excess of 7–10 mM have been reported to occur in humans following ingestion of methanol (McMartin et al., 1980; Sejersted et al., 1983).

Dorman et al. (1995) exposed pregnant CD-1 mice to methanol by inhalation at teratogenic concentrations (10,000 or 15,000 ppm for 6 hours on GD 8) and reported no significant increase in blood or decidual formate, indicating that formate is probably not involved in methanol teratogenesis in the mouse. Dosing of pregnant mice by gavage with Na^+-formate (750 mg/kg) was neither teratogenic nor raised maternal blood formate level. However, confirming the results of Andrews et al. (1993a, 1995), exposure of mouse embryos to either methanol (250 mM) or Na^+-formate (40 mM) in culture was developmentally toxic.

From the studies of formate in rodents *in vivo* and *in vitro* discussed above (Andrews et al., 1993b, 1995; Brown-Woodman et al., 1995; Dorman et al., 1995), it appears that formate is toxic to the developing embryo only at concentrations well in excess of those achieved after teratogenic methanol exposure in the mouse, but within the range of levels achieved in humans after acute high-dose methanol poisoning. Thus, although methanol appears to be the proximate murine teratogen, formate is still of concern in terms of potential developmental toxicity in humans.

A series of whole embryo culture studies by Harris and coworkers (Harris et al., 2003, 2004; Hansen et al., 2005) was designed to evaluate the role of methanol and its metabolites in methanol-induced developmental toxicity. Harris et al. (2004) showed that among methanol and its metabolites, viability of cultured rodent embryos is most affected by formate, while teratogenic endpoints were more sensitive to methanol and formaldehyde. Results from these studies suggest that developmental toxicity may be related to formaldehyde, as formaldehyde-induced teratogenicity occurs at several orders of magnitude lower than methanol (Hansen et al., 2005; Harris et al., 2004). It should also be noted that catalase, alcohol dehydrogenase (ADH) 1, and ADH3 activities are present in the rat embryo and visceral yolk sac (VYS) as early as 6–12 somites, which is during the early part of the culture period (Harris et al., 2003). Thus, it is plausible that, during this susceptible stage of development, methanol is metabolized to formaldehyde locally in the embryo and VYS.

Studies of embryonal glutathione (GSH) support the idea that formaldehyde may be a proximal teratogen following exposure to methanol. Inhibition of GSH synthesis with butathione sulfoximine (BSO) had little effect on the rat embryo developing *in vitro*, yet treatment with BSO and methanol or formaldehyde increased developmental toxicity (Harris et al., 2004). Among the enzymes involved in methanol clearance, only ADH3-mediated metabolism of formaldehyde is GSH-dependent. Interestingly, ADH3 activity is lower in mouse compared to rat embryos, which is consistent with the greater sensitivity of the mouse to methanol developmental toxicity (Andrews et al., 1993a; Harris et al., 2003; Hansen et al., 2005).

Studies such as those by Andrews et al. (1993a), Dorman et al. (1995, 1994), Harris et al. (2003, 2004), and Hansen et al. (2005) strongly suggest that formate is not the metabolite responsible for methanol's teratogenic effects. Harris and colleagues provide evidence that formaldehyde could be the proximate teratogen. However, research in this area is insufficient to conclusively identify formaldehyde as the proximate teratogen. Even if formaldehyde is ultimately identified as the proximate teratogen, methanol would likely play a prominent role, at least in transport to the target tissue. The high reactivity of formaldehyde would limit its transport as free formaldehyde from maternal to fetal blood (Thrasher and Kilburn, 2001), and the capacity for the metabolism of methanol to formaldehyde is likely lower in the embryo and fetus versus adults.

5.4 CONCLUSIONS

One of the most sensitive toxic effects of methanol when administered to experimental animals is the induction of developmental abnormalities in fetuses exposed *in utero*. Developmental effects have been demonstrated in a number of species, including rats, mice, and monkeys. Comparing the NOAELs and LOAELs for developmental toxicity in mice and rats similarly exposed *in utero* (e.g., Rogers et al., 1993a vs

Nelson et al., 1985) suggests that mice are the more sensitive species. Supporting evidence comes from *in vitro* studies in which rat and mouse embryos were exposed to methanol in culture (Andrews et al., 1993a). In the study of Rogers et al. (1993) in which pregnant female CD-1 mice were exposed to methanol vapors on GD 6–15 at a range of concentrations, fetal effects included an increase in resorptions, fewer live pups, and increased incidence of exencephaly, cleft palate, and cervical ribs. The cervical rib effect was the most sensitive indicator of developmental toxicity from this study, with a NOAEL of 1000 ppm. The gastrulation stage of mouse development was found to be the most sensitive to methanol (Rogers and Mole, 1997).

In rats, however, the most sensitive developmental effect, as reported in the NEDO (1987) two-generation inhalation studies, was lower brain weight at 3, 6, and 8 weeks postnatally after exposure of dams and their litters to 1000 ppm methanol by inhalation during gestation and throughout lactation. The NOAEL reported in this study was 500 ppm. No such evaluations of the effects of methanol on brain growth have been done in mice.

In studies by Burbacher and coworkers (Burbacher et al., 1999a, b, 2004a, b), exposure of monkeys (*M. fascicularis*) to methanol during premating, mating, and throughout gestation resulted in a shorter period of gestation in dams exposed to as low as 200 ppm. The shortened gestation in methanol-exposed groups was largely the result of C-sections due to maternal distress. Interpretation of these findings is complicated because gestational age, birth weight, and infant size in all groups were within normal ranges for *M. fascicularis*. There was also evidence that gestational methanol exposure caused neurobehavioral effects in offspring. However, the findings were not conclusive and the dose–response was not robust. Thus, there is insufficient evidence to determine if the primate fetus is more or less sensitive than rodents to the developmental toxicity of methanol.

Among the findings of the studies reviewed in this chapter that demonstrate the teratogenicity and developmental neurotoxicity of methanol, an increase in the incidence of cervical ribs of gestationally exposed mice at 2000 ppm methanol (Rogers et al., 1993a) and a decrease in the brain weights of gestationally and lactationally exposed rats at 1000 ppm methanol (NEDO, 1987) appear to be the most robust and most sensitive effects.

While it is established that the toxic consequences of acute methanol poisoning in adult humans arise from the action of formate, there is less certainty on how the toxicological impacts of long-term exposure to lower levels of methanol are brought about. Experiments carried out by Dorman et al. (1995) indicate that formate is not the proximate teratogen in pregnant CD-1 mice exposed to high concentrations of methanol vapor. Little, if any, formate accumulated in the blood of mice exposed to teratogenic concentrations of methanol, and treatment of pregnant mice with a high oral dose of formate did not induce neural tube closure defects. Finally, methanol, but not formate, induced neural tube closure defects in mouse embryos *in vitro* at concentrations comparable to the levels of methanol and formate detected in blood after a teratogenic exposure. Harris and colleagues (Hansen et al., 2005; Harris et al., 2003, 2004) concluded that the metabolite responsible for the developmental toxicity of methanol may be formaldehyde rather than the parent compound or formate. It was suggested that species differences in the metabolism of methanol to formaldehyde and formic acid by ADH1, ADH3, and catalase may contribute to differences in species sensitivity to methanol teratogenicity. The finding that ADH3 activity (converting formaldehyde to formate) was lower in the mouse VYS, and that inhibition of GSH synthesis increases the developmental toxicity of methanol, are consistent with the greater sensitivity of the mouse to the developmental toxicity of methanol. That formaldehyde may be the proximate teratogen is further supported by the finding that formaldehyde is embryotoxic to rat and mouse embryos *in vitro* at much lower concentrations than is either methanol or formate (Hansen et al., 2005).

While studies such as those by Harris et al. (Harris et al., 2003, 2004), Hansen et al. (2005), and Dorman et al. (1994, 1995) strongly suggest that formate is not the metabolite responsible for methanol's teratogenic effects, there are still questions regarding the relative involvement of methanol versus formaldehyde.

For further discussion of the role of metabolism in the teratogenicity of methanol, see chapter 7.

DISCLAIMER

The views expressed in this chapter are those of the authors and do not necessarily reflect the views or policies of the U.S. Environmental Protection Agency.

REFERENCES

Abbott, B.D., Ebron-McCoy, M., and Andrews, J.E. (1995) Cell death in rat and mouse embryos exposed to methanol in whole embryo culture. *Toxicology* **97**, 159–171.

Abbott B.D., Logsdon, T.R., and Wilke, T.S. (1994) Effects of methanol on embryonic mouse palate in serum-free organ culture. *Teratology* **49**, 122–134.

Andrews, J.E., Ebron-McCoy, M., Kavlock, R.J., and Rogers, J.M. (1995) Developmental toxicity of formate and formic acid in whole embryo culture: a comparative study with mouse and rat embryos. *Teratology* **51**, 243–251.

Andrews, J.E., Ebron-McCoy, M., Logsdon, T.R., Mole, L.M., Kavlock, R.J., and Rogers, J.M. (1993a) Developmental toxicity of methanol in whole embryo culture: a comparative study with mouse and rat embryos. *Toxicology* **81**, 205–215.

Andrews, J.E., Ebron-McCoy, M., Kavlock, R.J., and Rogers, J.M. (1993b) Lowering pH increases embryonic sensitivity to formate in whole embryo culture. *Toxicol. in Vitro* **6**, 757–762.

Antony A.C. (2007) In utero physiology: role of folic acid in nutrient delivery and fetal development. *Am. J. Clin. Nutr.* **85**, 598S–603S.

Aziz, M.H., Agrawal, A.K., Adhami, V.M., Ali, M.M., Baig, M.A., and Seth, P.K. (2002) Methanol-induced neurotoxicity in pups exposed during lactation through mother: role of folic acid. *Neurotoxicol. Teratol.* **24**, 519–527.

Bolon, B., Dorman, D.C., Janszen, D., Morgan, K.T., and Welsch, F. (1993) Phase-specific developmental toxicity in mice following maternal methanol inhalation. *Fundam. Appl. Toxicol.* **21**, 508–516.

Bolon, B., Welsch, F., and Morgan, K.T. (1994) Methanol-induced neural tube defects in mice: pathogenesis during neurulation. *Teratology* **49**, 497–517.

Brown-Woodman, P.D., Huq, F., Hayes, L., Herlihy, C., Picker, K., and Webster, W.S. (1995) In vitro assessment of the effect of methanol and the metabolite, formic acid, on embryonic development of the rat. *Teratology* **52**, 233–243.

Burbacher, T.M., Grant, K.S., Shen, D.D., Sheppard, L., Damian, D., and Ellis, S., et al., (2004a) Chronic maternal methanol inhalation in nonhuman primates (*Macaca fascicularis*): reproductive performance and birth outcome. *Neurotoxicol. Teratol.* **26**, 639–650.

Burbacher, T.M., Shen, D.D., Lalovic, B., Grant, K.S., Sheppard, L., and Damian, D., et al. (2004b) Chronic maternal methanol inhalation in nonhuman primates (*Macaca fascicularis*): exposure and toxicokinetics prior to and during pregnancy. *Neurotoxicol Teratol.* **26**, 201–21.

Burbacher, T., Shen D., Grant, K., Sheppard, L., Damian, D., Ellis, E., and Liberato, N. (1999a) Reproductive and offspring developmental effects following maternal inhalation exposure to methanol in nonhuman primates. Part I: Methanol distribution and reproductive toxicity in adult females. *Res. Rep. Health Eff. Inst.* **1999** (89), 1–68. Commentary 119–134.

Burbacher, T., Grant, K., Shen, D., Damian, D., Ellis, S., and Liberato, N. (1999b) Reproductive and offspring developmental effects following maternal inhalation exposure to methanol in nonhuman primates. Part II: Developmental effects in infants exposed prenatally to methanol. *Res. Rep. Health Eff. Inst.* **1999** (89), 69–117. Commentary 119–134.

Cameron, A.M., Nilsen, O.G., Haug, E., and Eik-Nes, K.B. (1984) Circulating concentrations of testosterone, luteinizing hormone and follicle stimulating hormone in male rats after inhalation of methanol. *Arch. Toxicol. Suppl.* **7**, 441–443.

Cameron, A.M., Zahlsen, K., Haug, E., Nilsen, O.G., and Eik-Nes, K.B. (1985) Circulating steroids in male rats following inhalation of n-alcohols. *Arch. Toxicol. Suppl.* **8**, 422–424.

Connelly, L.E. and Rogers, J.M. (1997) Methanol causes posteriorization of cervical vertebrae in mice. *Teratology* **55**, 138–144.

Cooper, R.L., Mole, M.L., Rehnberg, G.L., Goldman, J.M., McElroy, W.K., and Hein, J., et al. (1992) Effect of inhaled methanol on pituitary and testicular hormones in chamber acclimated and non-acclimated rats. *Toxicology* **71**, 69–81.

Cummings, A.M. (1993) Evaluation of the effects of methanol during early pregnancy in the rat. *Toxicology* **79**, 205–214.

Degitz, S.J., Rogers, J.M., Zucker, R.M., and Hunter, E.S., 3rd. (2004a) Developmental toxicity of methanol: pathogenesis in CD-1 and C57BL/6J mice exposed in whole embryo culture. *Birth Defects Res. A Clin. Mol. Teratol.* **70**, 179–184.

Degitz, S.J., Zucker, R.M., Kawanishi, C.Y., Massenburg, G.S., and Rogers, J. M. (2004b) Pathogenesis of methanol-induced craniofacial defects in C57BL/6J mice. *Birth Defects Res. A Clin. Mol. Teratol.* **70**, 172–178.

Dorman, D.C., Bolon, B., Struve, M.F., LaPerle, K.M., Wong, B.A., and Elswick, B., et al. (1995) Role of formate in methanol-induced exencephaly in CD-1 mice. *Teratology* **52**, 30–40.

Fagan, J.F. and Singer, L.T. (1983) Infant recognition memory as a measure of intelligence. In L.P. Lipsitt (Ed.), *Advances in Infancy Research* (Vol. 2, pp 31–78). Norwood, NJ: Ablex.

Fallang, B., Saugstad, O.D., Grogaard, J., and Hadders-Algra, M. (2003) Kinematic quality of reaching movements in preterm infants. *Pediatr. Res.* **53**, 836–842.

Fu, SS., Sakanashi, T.M., Rogers, J.M., Hong, K.H., and Keen, C.L. (1996) Influence of dietary folic acid on the developmental toxicity of methanol and the frequency of chromosomal breakage in the CD-1 mouse. *Reprod. Toxicol.* **10**, 455–463.

Hansen, J.M., Contreras, K.M., and Harris, C. (2005) Methanol, formaldehyde, and sodium formate exposure in rat and mouse conceptuses: a potential role of the visceral yolk sac in embryotoxicity. *Birth Defects Res. A Clin. Mol. Teratol.* **73**, 72–82.

Harris, C., Dixon, M., and Hansen, J.M. (2004) Glutathione depletion modulates methanol, formaldehyde and formate toxicity in cultured rat conceptuses. *Cell Biol. Toxicol.* **20**, 133–145.

Harris, C., Wang, S.W., Lauchu, J.J., and Hansen, J.M. (2003) Methanol metabolism and embryotoxicity in rat and mouse conceptuses: comparisons of alcohol dehydrogenase (ADH1), formaldehyde dehydrogenase (ADH3), and catalase. *Reprod. Toxicol.* **17**, 349–357.

Infurna, R. and Weiss, B. (1986) Neonatal behavioral toxicity in rats following prenatal exposure to methanol. *Teratology* **33**, 259–265.

Kavet, R. and Nauss, K.M. (1990) The toxicity of inhaled methanol vapors. *Crit. Rev. Toxicol.* **21**, 21–50.

Lee, E., Brady, A.N., Brabec, M.J., and Fabel, T. (1991) Effects of methanol vapors on testosterone production and testis morphology in rats. *Toxicol. Ind. Health.* **7**, 261–275.

McMartin, K.E., Ambre, J.J., and Tephly, T.R. (1980) Methanol poisoning in human subjects. Role for formic acid accumulation in the metabolic acidosis *Am. J. Med.* **68**, 414–418.

Medinsky, M.A., Dorman, D.C., Bond, J.A., Moss, O.R., Janszen, D.B., and Everitt, J.I. (1997) Pharmacokinetics of methanol and formate in female cynomolgus monkeys exposed to methanol vapors. *Res. Rep. Health Eff. Inst.* 1–30, discussion 31-8.

New Energy Development Organization (NEDO) (1987) Toxicological research of methanol as a fuel for power station. Summary report on tests with monkeys, rats and mice. New Energy Development Organization, Tokyo, Japan. Unpublished report.

Nelson, B.K., Brightwell, W.S., MacKenzie, D.R., Khan, A., Burg, J.R., and Weigel, W.W., et al., (1985) Teratological assessment of methanol and ethanol at high inhalation levels in rats. *Fundam. Appl. Toxicol.* **5**, 727–736.

Plantinga, Y., Perdock, J., and de Groot, L. (1997) Hand function in low-risk preterm infants: its relation to muscle power regulation. *Dev. Med. Child. Neurol.* **39**, 6–11.

Rogers, J.M., Brannen, K.C., Barbee, B.D., Zucker, R.M., and Degitz, S.J. (2004) Methanol exposure during gastrulation causes holoprosencephaly, facial dysgenesis, and cervical vertebral malformations in C57BL/6J mice. *Birth Defects Res. B Dev. Reprod. Toxicol.* **71**, 80–88.

Rogers, J.M. and Mole, M.L. (1997) Critical periods of sensitivity to the developmental toxicity of inhaled methanol in the CD-1 mouse. *Teratology* **55**, 364–372.

Rogers, J.M., Mole, M.L., Chernoff, N., Barbee, B.D., Turner, C.I., and Logsdon, T.R., et al., (1993) The developmental toxicity of inhaled methanol in the CD-1 mouse, with quantitative dose–response modeling for estimation of benchmark doses. *Teratology* **47**, 175–188.

Sakanashi, T.M., Rogers, J.M., Fu, S.S., Connelly, L.E., and Keen, C.L. (1996) Influence of maternal folate status on the developmental toxicity of methanol in the CD-1 mouse. *Teratology* **54**, 198–206.

Sejersted, O.M., Jacobsen, D., Ovrebo, S., and Jansen, H. (1983) Formate concentrations in plasma from patients poisoned with methanol. *Acta Med. Scand.* **213**, 105–110.

Stanton, M.E., Crofton, K.M., Gray, L.E., Gordon, C.J., Boyes, W.K., and Mole, M.L., et al., (1995) Assessment of offspring development and behavior following gestational exposure to inhaled methanol in the rat. *Fundam. Appl. Toxicol.* **28**, 100–110.

Sulik, K.K. and Johnston, M.C. (1983) Sequence of developmental alterations following acute ethanol exposure in mice: craniofacial features of the fetal alcohol syndrome. *Am. J. Anat.* **166**, 257–269.

Sulik, K.K., Johnston, M.C., and Webb, M.A. (1981) Fetal alcohol syndrome: embryogenesis in a mouse model. *Science* **214**, 936–938.

Tephly, T.R. and McMartin, K.E. (1984) Methanol metabolism and toxicity. *Food Sci. Technol.* **12**, 111–140.

Thrasher, J.D. and Kilburn, K.H. (2001) Embryo toxicity and teratogenicity of formaldehyde. *Arch. Environ. Health.* **56**, 300–311.

Webster, W.S., Walsh, D.A., McEwen, S.E., and Lipson, A.H. (1983) Some teratogenic properties of ethanol and acetaldehyde in C57BL/6J mice: implications for the study of the fetal alcohol syndrome. *Teratology* **27**, 231–243.

Weiss, B., Stern, S., Soderholm, S.C., Cox, C., Sharma, A., Inglis, G.B., Preston, R., Balys, M., Reuhl, K.R., and Gelein, R. (1996) Developmental neurotoxicity of methanol exposure by inhalation in rats. *Res. Rep. Health Eff. Inst.* **1996** (73), 1–64.

Yasuda, M., Sato, T.J., and Sumida, H. (1991) Excencephalic mouse fetuses are resistant to cleft palate induction with 2,3,7,8-tetrachlorodibenzo-*p*-dioxin (TCDD). *Teratology* **43**, 445 (abstract).

6 Exploring Differences Between PBPK Models of Methanol Disposition in Mice and Humans: Important Lessons Learned

THOMAS B. STARR

TBS Associates, Raleigh, NC, USA

6.1 BACKGROUND

While the United States Environmental Protection Agency (USEPA) classifies methanol as a "threshold" pollutant, the Agency has yet to establish an inhalation Reference Concentration (RfC) for this important commodity chemical. A Reference Concentration is defined as "an estimate (with uncertainty spanning perhaps an order of magnitude) of a continuous inhalation exposure to the human population (including sensitive subgroups) that is likely to be without an appreciable risk of deleterious effects during a lifetime" (USEPA, 2012). USEPA has been in the process of revising its Integrated Risk Information System (IRIS) methanol listing for several years now, and this process will most likely culminate in specification of a methanol RfC. The possibility that the Agency will characterize methanol as a potential human carcinogen cannot be ruled out at this time, so extrapolated human cancer risks may be the determining factor in setting an inhalation RfC. However, the weight of scientific evidence points strongly toward noncancer developmental toxicity as the primary health concern for human exposure. Either way, knowledge of methanol uptake, distribution, and disposition

The Toxicology of Methanol, First Edition. Edited by John J. Clary.

within the body, as well as elimination from it, will likely play a critical role in linking the relevant toxicity data to methanol exposure in the context of quantitative dose–response modeling.

Inhaled methanol vapors are absorbed readily by both primates and rodents. The primary metabolic pathway is oxidation to formaldehyde by alcohol dehydrogenase (in primates) or catalase (in rodents), with further rapid metabolism to formic acid by formaldehyde dehydrogenase (Mannering and Parks, 1975; Makar and Tephly, 1977). Formic acid is then further oxidized to carbon dioxide by a folate-dependent enzymatic pathway. In humans, the latter step becomes rate limiting at sufficiently high blood methanol levels, leading to metabolic acidosis, with the classic symptoms of methanol poisoning, for example, visual impairment, blindness, and death, as sequelae (McMartin et al., 1975, 1977; Tephly et al., 1979; Noker and Tephly, 1980). In contrast, for rodents, it is the first metabolic step, namely, oxidation of methanol to formaldehyde by catalase, which is rate limiting at high blood methanol levels (Tephly et al., 1979). The result is disproportionately large increases in rodent blood methanol levels with increasing methanol exposure (Clary, 2003).

The National Toxicology Program's Center for the Evaluation of Risks to Human Reproduction (NCERHR) completed a comprehensive review of the reproductive and developmental toxicity of methanol (CERHR, 2002). The CERHR Expert Panel report concluded that "developmental toxicity was the most sensitive endpoint of concern with respect to evaluating the risk to reproduction posed by methanol exposure in humans," and it identified the Rogers et al. (1993) study as a "critical study for the assessment of developmental toxicity." This study found dose-related increases in the frequency of cervical rib malformations among mouse pups whose dams had been exposed to airborne methanol concentrations of 0, 1000, 2000, 5000, or 15,000 ppm for 7 hours per day during a 10-day temporal window comprising gestation days 6 through 15. Of particular interest is the fact that Rogers et al. measured maternal blood methanol concentrations in a subset of the pregnant dams at the end of their 7-hour exposure periods on gestation days 6, 10, and 15, so these internal exposure data permit a quantitative dose–response assessment of developmental toxicity using blood methanol as an alternative (and preferred) dose metric to that of the methanol concentration in the inhalation chambers.

The CERHR Expert Panel review concluded that "blood methanol concentrations provide a useful dosimetric for comparison of results among various studies." Indeed, the Expert Panel explicitly characterized its levels of concern regarding the potential toxicity of different airborne methanol exposures in terms of the corresponding methanol concentrations in circulating blood. The Panel's characterization of blood methanol concentration as the "preferred" dose metric singles out methanol pharmacokinetics as a critical factor that needs careful consideration in any quantitative risk assessment of potential adverse human health effects that might arise from methanol exposure.

Starr and Festa (2003) undertook such a quantitative assessment using the developmental toxicity and blood methanol data collected by Rogers et al. (1993) to first determine a blood methanol RfC, and then convert that blood methanol RfC to a corresponding inhalation RfC with a physiologically based pharmacokinetic (PBPK) model of human methanol disposition that had been developed previously by Bouchard et al. (2001). The maximum likelihood estimate of their proposed methanol inhalation RfC was 298 mg/m^3, with a lower 95% confidence bound of 135 mg/m^3.

Subsequent discussions with USEPA regarding this proposal revealed several Agency concerns. First, the Agency commented that steady state might not have been reached in the Rogers et al. (1993) study where the mouse inhalation exposures lasted for only 7 hours each day. In contrast, the Starr and Festa (2003) analysis assumed that steady state had been reached well prior to exposure termination on each day of exposure. Second, the Agency commented that area under the blood methanol concentration curve (AUC) might be preferred over blood methanol concentration as the dose metric for USEPA's risk assessment purposes, and AUC data were not collected during the Rogers et al. (1993) study. These concerns eventually led USEPA to award a contract to investigators from Battelle's Pacific Northwest National Laboratory to develop a new PBPK model capable of describing methanol disposition in both mice and humans in a variety of exposure situations.

The Battelle work was aimed, at least in part, at developing estimates of the maximum blood methanol concentration (C_{max}) achieved during either 7 or 24 (i.e., continuous) hours of inhalation exposure and estimates of 24-hour AUC, following either 7- or 24-hour inhalation exposures to methanol. The Battelle investigators concluded that the human equivalent concentration corresponding to the inhalation route No Observed Effect Level (NOEL) of 1000 ppm from the Rogers et al. (1993) study was either 461.2 ppm (600 mg/m^3), based on an AUC dose metric, or 659 ppm (857 mg/m^3), based on a C_{max} dose metric (Battelle, 2006). Using the same uncertainty factors of 3 (for possible pharmacodynamic differences between species) and 10 (for interindividual differences in sensitivity) employed by Starr and Festa (2003), these values led to methanol inhalation RfCs of 20 or 28.6 mg/m^3, respectively, that is, 10- to 15-folds lower than the Starr and Festa (2003) maximum likelihood estimate. This chapter describes a systematic exploration and comparison of various predictions made by the Battelle (2006) and Bouchard et al. (2001) pharmacokinetic models of methanol disposition. This comparison has yielded a number of insights that should prove useful in future pharmacokinetic modeling and quantitative risk assessment efforts for methanol, as well as for other volatile compounds.

6.2 ARE HUMANS MORE OR LESS SENSITIVE THAN MICE TO THE TOXIC EFFECTS OF METHANOL?

The Battelle (2006) report (Table 2 and Figure 13, data not shown) provides results from PBPK model simulations of mouse and human AUC and C_{max} as functions of airborne methanol concentration. Inspection of the AUC results revealed that the predicted human AUC always exceeds the corresponding mouse AUC. At 1 ppm, the difference is a factor of 3.62, but the predicted human to mouse AUC ratio grows to almost 16.4 at 1000 ppm. Then it declines gradually to 7.9 at 5000 ppm. Similarly, the human C_{max} always exceeds the mouse C_{max}. At 1 ppm, the difference is quite small, a factor of 1.06, but at 1000 ppm, the predicted human to mouse C_{max} ratio grows to 3.95 and then declines gradually to 2.86 at 5000 ppm.

However, the mouse results cited earlier pertain to exposure conditions in the Rogers et al. (1993) study, that is, C_{max} is the blood concentration immediately following 7 hours of exposure, and the AUC is computed for a 24-hour period starting at the onset of a 7-hour exposure period. In contrast, the human results pertain to continuous 24-hour exposure, so C_{max} is the predicted blood methanol concentration at steady state, and the 24-hour AUC is just $24 \times C_{max}$. Thus, the comparison of mouse and human C_{max} and AUC values as provided in the Battelle (2006) report is confounded by the different durations of the mouse and human exposure regimens. Indeed, it could be argued that the different exposure durations (7 hours for mice, 24 hours for humans) are responsible for most of the apparent discrepancy between thc mouse and human C_{max} and AUC values.

To resolve this species sensitivity question, an "apples to apples" comparison of the two species responses is required. Since lifetime continuous exposure at the fence line is the primary concern of the inhalation RfC process, a comparison of human and mouse C_{max} and 24-hour AUC is most appropriately undertaken for continuous exposure conditions.

This was accomplished by solving (algebraically) the differential equations representing the Battelle PBPK model under steady-state exposure conditions.

At steady state, the Battelle PBPK model reduces to the following equations, using the notation employed in Appendix B.1 of the Battelle (2006) report:

Arterial Blood:	$\mathrm{CVLu} = \mathrm{CAB}$
Fat:	$\mathrm{CVF} = \mathrm{CAB}$
Rest of Body:	$\mathrm{CVR} = \mathrm{CAB}$
Liver:	$\mathrm{CAB} = \mathrm{CVL}^*(1 + \mathrm{V}_{max}/(\mathrm{QL}^*(\mathrm{KM} + \mathrm{CVL})) + \mathrm{V}_{max}2/(\mathrm{QL}^*(\mathrm{KM2} + \mathrm{CVL})))$
Mixed Venous Blood:	$(\mathrm{QC} + \mathrm{KL}^*\mathrm{VVB})^*\mathrm{CVB} = (\mathrm{QF} + \mathrm{QR})^*\mathrm{CAB} + \mathrm{QL}^*\mathrm{CVL}$
Lungs:	$\mathrm{QP}^*(\mathrm{FRACin}^*\mathrm{CCh} - \mathrm{CVB}/\mathrm{PB}) = \mathrm{QC}^*(\mathrm{CAB} - \mathrm{CVB})$

These nonlinear equations can be manipulated so as to express the steady-state arterial and mixed venous blood methanol concentrations

(CAB and CVB, respectively) as well as the airborne methanol concentration (CCh) entirely in terms of fixed PBPK model parameters and the steady-state concentration of methanol in the venous blood leaving the liver (CVL). At steady state, the latter quantity is the same as C_{max} in the Battelle (2006) model, and $AUC = 24 \times C_{max}$. Thus, any inferences that can be drawn regarding species sensitivity differences in C_{max} apply equally to 24-hour AUC at steady state.

The final human and mouse parameter values from the "visual optimization" employed by Battelle are reproduced in Table 6.1 (see "Mouse" and "Human" columns). These values were utilized in computing steady-state mouse and human venous blood methanol concentrations with the steady-state PBPK model equations described earlier. Results of these computations are presented in tabular form (Table 6.2) and graphically in Figure 6.1. Also shown is

TABLE 6.1 Battelle (2006) PBPK Model Parameter Values. Human Metabolism Parameters That were Changed in the Modified Battelle Human Model are Italicized

Model Parameter	Mouse	Human	Modified Human
bw, kg	0.03	70	70
$bw^{0.75}$	0.072084	24.20045	24.20045
V_{max}, mg/h	1.030806	266.205	*6980.156*
V_{max2}, mg/h	1.369603	0	0
K_m, mg/l	12	12	*460*
K_{m2}, mg/l	210	0	0
QC, l/h	1.830942	447.7084	447.7084
QL, l/h	0.457736	101.6298	101.6298
QF, l/h	0.091547	23.28084	23.28084
QR, l/h	1.28166	322.7978	322.7978
kL, /h	0.	0.015212	0.015212
VVB, kg	0.001104	4.151	4.151
QP, l/h	1.830942	447.7084	447.7084
Fracin	0.73	0.75	0.75
PB	1350	1626	1626

TABLE 6.2 Mouse and Human Blood Methanol Concentrations (mg/l) Predicted by the Battelle (2006) and Bouchard et al. (2001) Models at Steady State for Various Airborne Methanol Concentrations (ppm)

Methanol in Air, (ppm)	Battelle Mouse, (mg/l)	Battelle Human, (mg/l)	Ratio: Bat H / Bat M	Bouchard Human, (mg/l)	Ratio: Bat H / Bou H
1	0.0213	0.0226	1.060	0.0313	0.722
10	0.216	0.229	1.063	0.313	0.732
50	1.134	1.213	1.069	1.565	0.775
100	2.429	2.622	1.079	3.130	0.838
250	7.685	8.766	1.141	7.825	1.12
500	25.78	41.17	1.597	15.65	2.63
525	28.71	49.13	1.711	16.43	2.99
550	31.91	56.98	1.786	17.22	3.31
575	35.4	71.07	2.008	18.00	3.95
600	39.22	85.69	2.185	18.78	4.56
625	43.36	103.0	2.374	19.56	5.26
675	52.71	144.9	2.749	21.13	6.96
750	69.56	221.5	3.184	23.48	9.44
875	106.0	367.0	3.463	27.39	13.4
1,000	153.9	520.9	3.385	31.30	16.6
2,000	980.3	1,798.0	1.834	62.60	28.7
5,000	4,681.2	5,663.3	1.210	156.5	36.2
10,000	11,061.3	12,110.4	1.095	313.0	38.7

Also shown are ratios of Battelle (2006) model-predicted human to Battelle (2006) model-predicted mouse blood methanol concentrations (Bat H / Bat M), and Battelle (2006) model-predicted human to Bouchard et al. (2001) model-predicted human blood methanol concentrations (Bat H / Bou H). See also Figures 6.1 and 6.2

the ratio of the Battelle (2006) model predicted steady-state human and mouse blood concentrations as a function of airborne methanol concentration.

At no airborne methanol concentration is the Battelle (2006) model-predicted steady-state human blood methanol concentration smaller than the corresponding mouse concentration, but the differences are less than about 10% below 250 ppm. However, above 250 ppm the

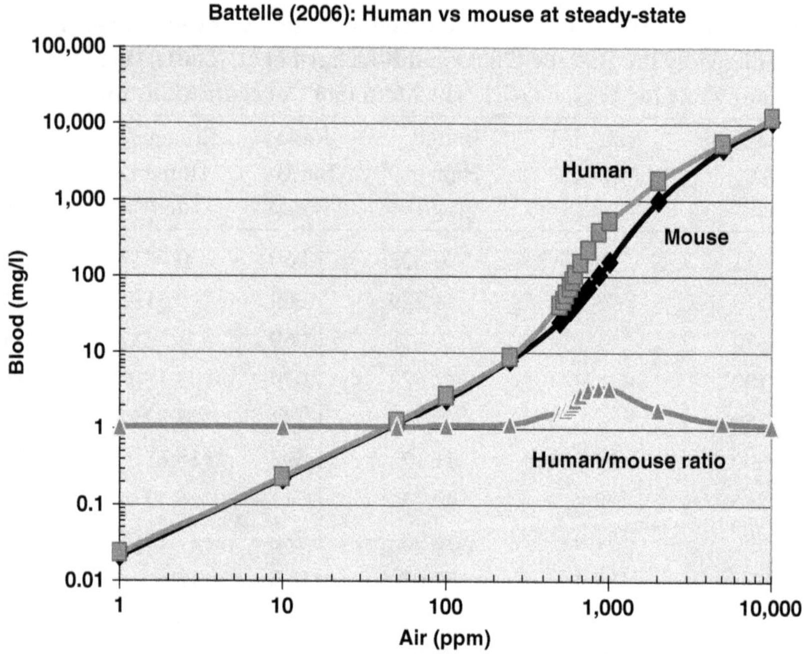

FIGURE 6.1 Battelle (2006) model-predicted mouse and human steady-state venous blood methanol concentrations versus airborne methanol concentration for continuous inhalation exposure. Also depicted is the ratio of predicted steady-state human and mouse blood concentrations.

Battelle (2006) model-predicted human and mouse steady-state values diverge markedly, and the maximum difference, an approximately 3.5-fold factor (367 mg/l vs 106 mg/l), is reached at 875 ppm, after which the ratio of Battelle (2006) model-predicted human to mouse blood concentrations begins to decline again. At 10,000 ppm and higher airborne methanol concentrations, the predicted differences are again less than 10%.

The present analysis thus implies that the Battelle (2006) model-predicted human steady-state blood methanol concentrations are always greater than the corresponding mouse concentrations, with the greatest differences occurring in the vicinity of 1000 ppm, the Rogers et al. (1993) NOEL level, where the Battelle model-predicted human steady-state concentration is about 3.4-fold higher than the corresponding mouse concentration (521 mg/l vs 154 mg/l). Assuming that there are no pharmacodynamic differences between mice and humans, these Battelle (2006) model-predicted results imply that humans are actually more, not less, sensitive than mice when both are exposed continuously to the same airborne methanol concentration. This implication stands in direct contradiction to the extensive body of evidence in the scientific literature indicating that the mouse is uniquely sensitive, relative to all other species tested, to the toxic effects of methanol, due principally to its limited metabolic capability for methanol (CERHR, 2002; Clary, 2003).

6.3 ARE THE TWO MODELS' PREDICTIONS OF HUMAN BLOOD METHANOL CONCENTRATIONS AT STEADY STATE CONSISTENT WITH EACH OTHER?

The Bouchard et al. (2001) model is a multicompartment, biologically based, dynamic model that describes the time evolution of methanol and its metabolites in the whole body, excreted urine, and expired air. A single system of first-order differential equations with species-specific parameter values for rats, monkeys, and humans accounts for the uptake, distribution, and metabolic conversion of methanol into formaldehyde, formate, and carbon dioxide. Bouchard et al. determined human parameter values by fitting the model outputs to blood and urinary methanol concentration time-course data obtained by other investigators using either cynomolgus monkeys (Dorman et al., 1994) or human volunteers (Osterloh et al., 1996; Sedivec et al., 1981). They then validated the human version of the model with completely independent methanol time-course data from human volunteer inhalation exposures as reported by Batterman et al. (1998).

A key feature of the Bouchard et al. (2001) model is that the kinetics of methanol in humans and monkeys are linear throughout the concentration ranges where human data are available. Only the rat version of this model required saturable Michaelis–Menten metabolism, and that was required only for the first metabolic step, namely, conversion of methanol to formaldehyde by catalase. An important consequence of linear kinetics in the human volunteer exposure range (up to 800 ppm) is that the increment in blood methanol concentration resulting from inhalation exposure is directly proportional to the airborne methanol concentration at every instant although the proportionality constant is time dependent until steady state is reached. Starr and Festa (2003) solved the Bouchard et al. (2001) human methanol model equations at steady state and determined that the relationship between airborne (C_{Air}, in ppm) and incremental blood (C_{max}, in mg/l) methanol concentrations at steady state was as follows:

$$C_{max} = 0.0313 \times C_{Air}.$$

Table 6.2 also presents the steady-state human blood methanol concentrations predicted by the Bouchard et al. (2001) model for comparison with those predicted by the Battelle (2006) model.

This comparison is also depicted graphically in Figure 6.2. While the Battelle (2006) and Bouchard et al. (2001) predictions of human blood methanol concentrations are in good agreement for airborne methanol

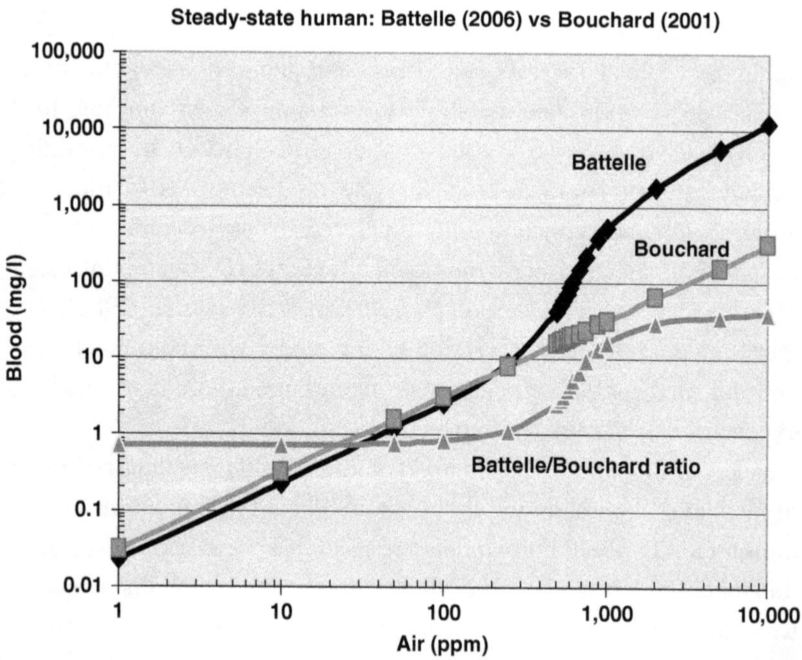

FIGURE 6.2 Battelle (2006) and Bouchard et al. (2001) model-predicted steady-state human blood concentrations during continuous (steady-state) exposure to various airborne methanol concentrations. Also depicted is the ratio of these two predicted concentrations.

concentrations of 250 ppm or below, they are *grossly inconsistent* at higher airborne methanol concentrations. At 500 ppm, the Battelle (2006) model predicts a human blood methanol concentration of 41.2 mg/l, while the Bouchard et al. (2001) model predicts 15.7 mg/l, nearly threefold lower. At 1000 ppm, the Battelle (2006) model predicts 521 mg/l, while the Bouchard et al. (2001) model predicts 31.3 mg/l, nearly 17-fold lower. At 10,000 ppm, the Battelle (2006) model predicts 12,110 mg/l, while the Bouchard et al. (2001) model predicts only 313 mg/l, nearly 39-fold lower. As is discussed in greater detail below, these gross inconsistencies can be traced to flawed estimates of key metabolic parameters in the Battelle (2006) human methanol model.

6.4 ARE THE VALUES OF KEY HUMAN METABOLISM PARAMETERS CONSISTENT WITH THOSE IN THE PUBLISHED SCIENTIFIC LITERATURE?

The Battelle (2006) model describes methanol metabolism with saturable Michaelis–Menten kinetics in both mice and humans. However, while the mouse model has two saturable metabolic pathways, one with a saturation constant of 12 mg/l (K_m), the other with a saturation constant of 210 mg/l (K_{m2}), the human has just one saturable metabolic pathway with a saturation constant exactly equal, surprisingly, to the first mouse pathway value, namely, 12 mg/l. No rationale was supplied for why the final human parameter estimate of K_m is exactly the same as the final mouse K_m estimate. Apparently this was just assumed to be the case. This human value is surprising because it implies that human methanol metabolism is already halfway to saturation at the comparatively low venous blood concentration exiting from the liver of 12 mg/l, a prediction that contradicts considerable evidence in the published scientific literature.

For example, when Bouchard et al. (2001) calibrated and then validated their model of methanol disposition in rats, monkeys, and humans, a saturation constant was required only to describe methanol's metabolism in rats. In monkeys and humans, the metabolism was simply described with first-order kinetics, that is, completely without saturation, and there was excellent agreement between the Bouchard et al. (2001) model predictions and measured blood methanol levels up to nearly 100 mg/l in cynomolgus monkeys exposed via inhalation to 900 ppm for 2 hours (Dorman et al., 1994), in rhesus monkeys exposed via inhalation to methanol concentrations of either 1200 or 2000 ppm for 6 hours, and up to nearly 14 mg/l in humans exposed via inhalation to 800 ppm for 2 hours (Batterman et al., 1998).

Indeed, none of the human data utilized in estimating the parameters of the Battelle (2006) human model demonstrate evidence of saturation of methanol metabolism, including the most recently published report relied on by Battelle, that of Ernstgard et al. (2005). These latter investigators clearly stated that their study "indicates nonsaturated, dose-proportional kinetics of methanol up to 200 ppm for 2 hours" (200 ppm was the highest airborne methanol concentration that was tested in this study). Ernstgard et al. (2005) stated that linear kinetics were supported by an average K_m estimate of 460 mg/l for human and monkey alcohol dehydrogenases reported by Perkins et al. (1995). This K_m value was over 38-fold larger than the 12 mg/l value utilized in the Battelle (2006) model. In fact, the only specific human estimate of K_m alluded to by Perkins et al. (1995) was 640 mg/l, as previously reported in Pirola (1978), over 53-fold larger than the Battelle (2006) model value. Saturation of the human metabolism of methanol thus seemed to be achieved only after severe, life-threatening oral overdoses (Jacobsen et al., 1988).

Misspecification of K_m can have an enormous impact on the Battelle (2006) model's predictions of human blood methanol levels arising from inhalation exposures, especially to high methanol concentrations, and this is easily demonstrated. If (1) the Battelle (2006) human K_m estimate of 12 mg/l is replaced by 460 mg/l, the average value reported in Perkins et al. (1995) and (2) the corresponding human V_{max} estimate is adjusted upward to give the same low-dose slope for the blood methanol (CVB) versus airborne concentration (CCh) dose–response relationship as that of the Bouchard et al. (2001) model at steady state, namely, 0.0313 mg/l/ppm, then this modified Battelle model-predicted human dose–response relationship is essentially linear at concentrations of 5000 ppm and below. Substantial nonlinearity only sets in at levels approaching 10,000 ppm.

This is evident in Table 6.2, and also in Figure 6.3, where the steady-state Battelle (2006) model-predicted mouse and (modified) human blood concentrations are plotted versus airborne methanol concentration for direct comparison with the unmodified Battelle human model predictions presented previously in Table 6.2 and Figure 6.1. At 10,000 ppm, the unmodified Battelle (2006) model predicted a human blood methanol concentration of 12,110 mg/l, while the modified Battelle model predicts only 697 mg/l, over 38-fold lower. At

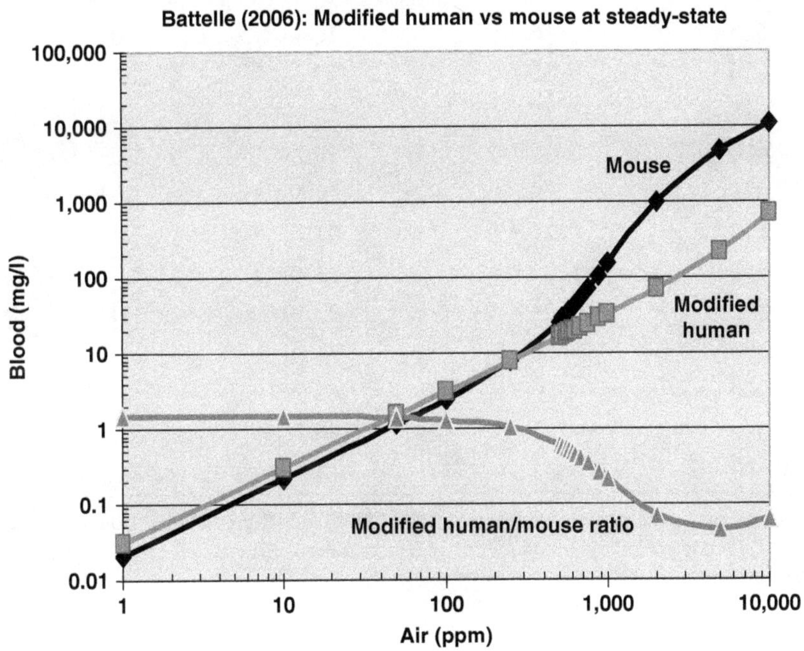

FIGURE 6.3 Battelle (2006) model-predicted mouse and human steady-state venous blood methanol concentrations versus airborne methanol concentration for continuous inhalation exposure. Also depicted is the ratio of predicted steady-state human and mouse blood concentrations. The human Battelle (2006) model was modified by (1) replacement of the human K_m estimate of 12 mg/l used by Battelle with the average value of 460 mg/l for humans and monkeys reported by Perkins et al. (1995) and (2) adjustment of the corresponding human V_{max} to give the same low-dose slope for the blood methanol (CVB) versus airborne concentration (CCh) relationship as that of the Bouchard et al. (2001) model at steady state.

TABLE 6.3 Human Blood Methanol Concentrations (mg/l) Predicted by the Modified Battelle (2006) and Bouchard et al. (2001) Models at Steady State for Various Airborne Methanol Concentrations (ppm)

Methanol in Air, ppm	Modified Battelle Human, mg/l	Bouchard Human, mg/l	Ratio: ModBat H / Bou H
1	0.0313	0.0313	1.000000
10	0.313	0.313	1.000639
50	1.569	1.565	1.00281
100	3.147	3.130	1.00550
250	7.932	7.825	1.0137
500	16.08	15.65	1.0276
525	16.9	16.43	1.0291
550	17.7	17.22	1.0304
575	18.6	18.00	1.0319
600	19.4	18.78	1.0333
625	20.2	19.56	1.0346
675	21.9	21.13	1.0375
750	24.5	23.49	1.0420
875	28.7	27.39	1.0494
1,000	33.1	31.30	1.0569
2,000	70.2	62.60	1.1209
5,000	217.3	156.5	1.3883
10,000	697.5	313.0	2.2284

Also shown is the ratio of modified Battelle (2006) model-predicted to Bouchard et al. (2001) model-predicted blood methanol concentrations (ModBat H / Bou H). See also Figure 6.4

1000 ppm, the original Battelle model predicted a human blood methanol concentration of 521 mg/l, while the modified model predicts only 33 mg/l, nearly fivefold lower.

It is also important to note that the implications of the modified Battelle model regarding human sensitivity to the toxic effects of methanol exposure relative to that observed in mice are completely opposite to those of the original Battelle model. Rather than predicting humans to be more sensitive than mice, and especially so near to the Rogers et al. (1993) NOEL dose of 1000 ppm, the modified Battelle model predicts humans to be much less sensitive than mice, by a nearly fivefold factor (153.9 mg/l vs 33.1 mg/l) at 1000 ppm, and by a nearly 22-fold factor (4681.2 mg/l vs 217.3 mg/l) at 5000 ppm. At airborne methanol concentrations of 250 ppm or less, however, human and mouse blood methanol levels are very similar when both are continuously exposed.

Finally, Table 6.3 and Figure 6.4 provide a direct comparison between human steady-state blood concentrations predicted with the modified Battelle (2006) model and the Bouchard et al. (2001) model. The agreement between these two sets of model predictions is striking, with differences no larger than 6% over a wide 1000-fold airborne concentration range. It is only at the very highest concentrations (<1000 ppm) that the saturable kinetics of the modified Battelle (2006) model leads to noticeable differences from the straight-line

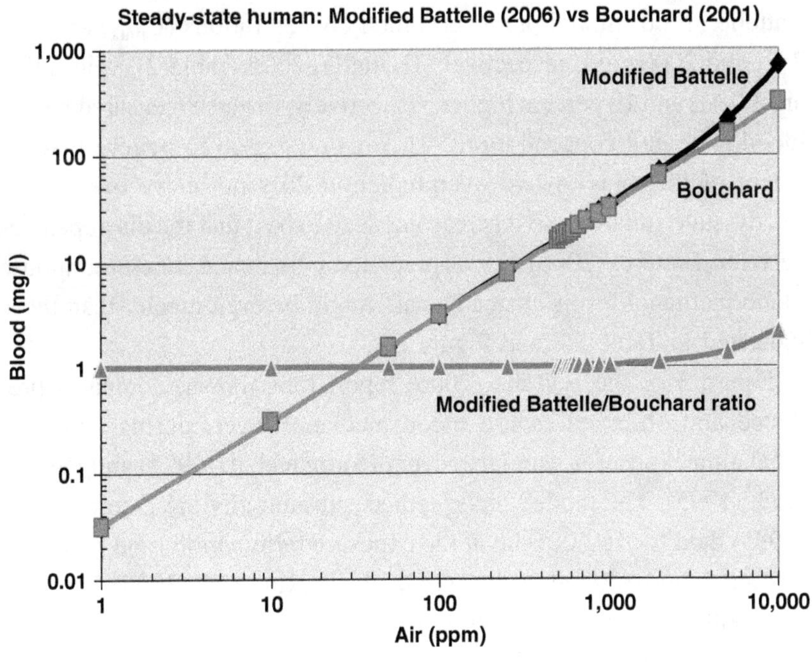

FIGURE 6.4 Modified Battelle (2006) and Bouchard et al. (2001) model-predicted steady-state human blood concentrations during continuous (steady-state) exposure to various airborne methanol concentrations. Also depicted is the ratio of these two predicted concentrations.

steady-state human blood concentrations generated with the linear kinetics of the Bouchard et al. (2001) model.

6.5 SHOULDN'T THE POSSIBILITY OF SYSTEMATIC BIAS BE CONSIDERED CAREFULLY DURING THE MODEL FITTING AND PARAMETER ESTIMATION PROCESS?

As noted previously, significant dose-related effects of methanol on the frequency of cervical rib malformations in mouse pups from pregnant dams exposed to methanol 7 hours/day via inhalation were observed in the Rogers et al. (1993) study. The key exposure levels were 1000, 2000, and 5000 ppm, and the grand mean measured blood methanol concentrations immediately after selected 7-hour exposure periods (days 6, 10, and 15) were 97 ± 16, 537 ± 77, and 1690 ± 160 mg/l (mean $\pm$ SEM), respectively (see Starr and Festa (2003), Table 1). The corresponding Battelle (2006) model-predicted blood concentrations (C_{max}) are 134, 757, and 2000 mg/l, respectively (Battelle, 2006, Table 2), values that are 38, 41, and 18 percent higher, respectively, than the measured mouse blood methanol concentrations. There is no reason to believe that the extent of this bias toward overprediction does not carry over to the steady-state situation, so it is reasonable to expect that the discrepancies between Battelle (2006) model-predicted human and measured mouse blood methanol levels at steady state might be even greater than those indicated in Table 6.2 and Figure 6.1.

Figure 3 of the Battelle (2006) report (not shown) compares predicted and observed mouse blood methanol levels during and after inhalation exposures using data from Dorman et al. (1995) and Perkins et al. (1995). The model underpredicts substantially the Dorman et al. (1995) data for 15,000 ppm at the exposure termination time point and beyond. It underpredicts substantially the Perkins et al. (1995) data from the 5000 ppm exposure group at all time points.

Figure 4 of the Battelle (2006) report (not shown) compares predicted and observed mouse blood methanol levels following oral exposures using data from Dorman et al. (1995) and Ward et al. (1997). The data for oral exposures of 2500 mg/kg are substantially underpredicted at all time points. Similarly, Figure 9 of the Battelle (2006) report (not shown) compares predicted and observed human blood methanol levels during and after inhalation exposures using data from Ernstgard et al. (2005). The data for the 100 ppm exposures are substantially underpredicted at almost all time points.

Figure 10 of the Battelle (2006) report (not shown) compares predicted and observed human blood methanol levels during and after 200 ppm inhalation exposures using data from Osterloh et al. (1996). The data are substantially underpredicted during exposure even after subtraction of the reported background concentrations. Similarly, Figure 11 of the Battelle (2006) report (not shown) compares predicted and observed human blood methanol levels during and after 800 ppm inhalation exposures using data from Batterman et al. (1998). The data for all time points are substantially overpredicted during exposure after subtraction of the reported background concentrations.

Given all of these instances of systematic bias of the Battelle (2006) model predictions relative to the data that were employed in developing and fitting it, perhaps it would be prudent to consider instead a sequential fitting process in which data sets are incorporated one at a time, thus permitting a determination of how the fit to previously included data sets is improved or degraded as additional data are included.

6.6 IS "VISUAL OPTIMIZATION" AN ADEQUATE TECHNIQUE FOR ESTIMATING PBPK MODEL PARAMETERS?

When PBPK models were first being developed, the software required for rigorous statistical parameter estimation was completely lacking. Initial values of many model parameters were often just "educated guesses," and parameter "optimization" was accomplished by trial and error, with visual inspection of discrepancies between data and predicted time courses providing the sole basis for determining whether a given model's "fit" to the data was acceptable. Clearly, this process involved a great deal of subjectivity, leaving numerous opportunities for the unwitting introduction of small, but systematic, and possibly critical, errors.

The most critical deficiency of the visual optimization technique employed by Battelle is readily apparent in Table 6.1 of the Battelle (2006) report, the summary final model parameter values: no standard errors of estimate are provided. It is thus impossible to quantify the bias, uncertainty, and/or variability associated with the Battelle (2006) model's final parameter estimates. Some estimates, such as those for body weights, may well be very precise and accurate, while others, such as the human K_m and V_{max} estimates, may have residual uncertainties and/or biases that span orders of magnitude. Visual optimization fails to provide such critical information regarding bias, uncertainty, and/or variability.

Importantly, the human inhalation data that were used in model calibration by Battelle show no evidence of the saturation of methanol metabolism. Thus, only the ratio of V_{max} to K_m will be well specified (c.f., Sweeney et al., 2004). In such circumstances, methanol metabolism is most efficiently represented by pseudo-first-order kinetics, as was employed by Bouchard et al. (2001). Other data, such as those cited in Perkins et al. (1995), are required to specify the human values of K_m and V_{max} individually. The Battelle (2006) report's final human K_m estimate of 12 mg/l is exactly the same as its mouse estimate, and, as noted previously, this is altogether inconsistent with the human and monkey values reported in the published scientific literature. This suggests strongly that the visual optimization technique employed by Battelle (2006) was simply incapable of generating reliable estimates of the key parameters (V_{max} and K_m) that together specify the human metabolism of methanol.

Faced with a bewildering array of data from multiple sources involving multiple routes of exposure, a more formal and systematic approach to parameter estimation than that provided by visual optimization is required, and the approach best suited to this complex task is Bayesian Markov chain Monte Carlo (MCMC) modeling (c.f., Gelman et al., 1996; Bernillon and Bois, 2000). With the MCMC approach, initial PBPK model parameter estimates, including estimated standard errors, are extracted from the peer-reviewed scientific literature. This information is then used to construct a prior multivariate probability distribution of the model's parameters, which is subsequently updated via fitting to (i.e., simulation of) additional experimental data that have not been used to specify the prior distribution, in a stepwise manner. As new data sets are added, one or more at a time, their contributions to parameter uncertainty/variability and bias reduction can be assessed objectively in terms of the extent to which resulting posterior mean parameter estimates shift away from their prior values and their posterior variance estimates shrink or widen.

The power of the MCMC approach in this Bayesian estimation context is its capacity to achieve convergence to final parameter values even when the PBPK model and its parameter covariance structure are extremely complex and the data available for model calibration are very sparse. Interestingly, USEPA is already quite familiar with the compelling advantages of the MCMC approach to PBPK model estimation since it recently (1) cosponsored a state-of-science monograph that included multiple applications of this approach to trichloroethylene (USEPA, 2000) and (2) published a new PBPK model for benzene with parameters estimated using MCMC methodology (Yokley et al., 2006).

Since the Rogers et al. (1993) data set is crucial to the methanol risk assessment process, particularly for the 1000, 2000, and 5000 ppm exposure groups, it is important that these data receive the greatest weight when other data sets are included in the fitting process. Indeed, it might well be appropriate to select the mouse prior parameter distribution to reproduce, as closely as possible, the Rogers et al. (1993) mean blood methanol levels 7 hours after exposure onset. Then, as other data sets (e.g., those from Dorman et al. (1995) and Perkins et al. (1995)) are brought into consideration, their contribution, if any, to bias and uncertainty/variability reduction can be systematically and objectively determined.

6.7 WHEN HUMAN DATA ARE AVAILABLE, SHOULDN'T THEY BE UTILIZED IN MAKING AN OBJECTIVE COMPARISON OF MODEL-SPECIFIC PREDICTIONS?

Many of the data sets employed by Battelle in human model parameter estimation were also employed by Bouchard et al. (2001) for either human model calibration or validation (c.f., Osterloh et al., 1996; Sedivic et al., 1981; Batterman et al., 1998). As noted previously, the Bouchard et al. (2001) model is more linear and simpler than the Battelle (2006) model, yet it accurately predicts human blood methanol levels arising from inhalation exposures up to the highest concentration (800 ppm) that has been tested. Given the excellent performance of the Bouchard et al. (2001) human model, it is not clear that the Battelle (2006) human model is in any way superior. An objective comparison of predictions by the Battelle (2006) and Bouchard et al. (2001) human models using human blood methanol data would go a long way toward answering this important question.

6.8 SUMMARY OF LESSONS LEARNED

Except for the first, each of the preceding sections of this chapter began by posing a critical question that was then addressed in considerable detail using the Battelle (2006) and Bouchard et al. (2001) models of methanol disposition in mice and humans in exemplary fashion. Here, the answers to these questions, which comprise our "lessons learned," are summarized:

- The mouse is uniquely sensitive, relative to all other species tested, to the toxic effects of methanol, due principally to its limited metabolic capability for methanol. If PBPK modeling suggests otherwise, trust the data; suspect the model.
- If model predictions are discordant with those of other validated models and/or experimental data, suspect the model and search for errors in its parameterization.
- Check to confirm that the numerical values assigned to model parameters are consistent with those in the published literature.
- During the model fitting and parameter estimation process, inspect simulation results carefully for the possibility of systematic bias. Attempt to discover its cause.
- "Visual optimization" should be abandoned as a model fitting technique; it should be replaced by a Bayesian Markov chain Monte Carlo (MCMC) modeling approach.

- When human data, especially independent human data, are available, they should always be utilized to objectively challenge model predictions.

REFERENCES

Battelle. (2006) Development, Calibration and Application of a Methanol PBPK Model. Battelle Project No. 48746. Authors: Poet, T.S., Teeguarden J.G., and Hinderliter, P.M. Final Report dated January 10, 2006. Pacific Northwest National Laboratory, Center for Biological Monitoring and Modeling, Richland, WA.

Batterman, S.A., Franzblau, A., D'Arcy, J.B., Sargent, N.E., Gross, K.B., and Schreck, R.M. (1998) Breath, urine, and blood measurements as biological exposure indices of short-term inhalation exposure to methanol. *Int. Arch. Occup. Environ. Health* **71**, 325–335.

Bernillon, P. and Bois, F.Y. (2000) Statistical issues in toxicokinetic modeling: A Bayesian perspective. *Environ. Health Perspect.* **108**, (Suppl 5), 883–893.

Bouchard, M., Brunet, R.C., Droz, P.O., and Carrier, G. (2001) A biologically based dynamic model for predicting the disposition of methanol and its metabolites in animals and humans. *Toxicol. Sci.* **64**, 169–184.

CERHR (2002) NTP-CERHR Expert Panel Report on the Reproductive and Developmental Toxicity of Methanol. NTP-CERHR-MeOH-02. April 2002. Center for the Evaluation of Risks to Human Reproduction, Research Triangle Park, NC.

Clary, J.J. (2003) Methanol, is it a developmental risk to humans? *Reg. Toxicol. Pharmacol.* **37**, 83–91.

Dorman, D.C., Moss, O.R., Farris, G.M., Janszen, D., Bond, J.A., and Medinsky, M.A. (1994) Pharmacokinetics of inhaled (^{14}C) methanol and methanol-derived ^{14}C formate in normal and folate-deficient cynomolgus monkeys. *Toxicol. Appl. Pharmacol.* **128**, 229–238.

Dorman, D.C., Bolon, B., Struve, M.F., LaPerle, K.M., Wong, B.A., Elswick, B., and Welsch, F. (1995) Role of formate in methanol-induced exencephaly in CD-1 mice. *Teratology* **52** (1), 30–40.

Ernstgard, L., Shibata, E., and Johanson, G. (2005) Uptake and disposition of inhaled methanol vapor in humans. *Toxicol. Sci.* **88** (1), 30–38.

Gelman, A., Bois, F., and Jiang, J. (1996) Physiological pharmacokinetic analysis using population modeling and informative prior distributions. *J. Am. Stat. Assoc.* **91**, 1400–1412.

Jacobsen, D., Webb, R., Collins, T.D., and McMartin, K.E. (1988) Methanol and formate kinetics in late diagnosed methanol intoxication. *Med. Toxicol.* **3**, 418–423.

Makar, A.B. and Tephly, T.R. (1977) Methanol poisoning. VI. Role of folic acid in the production of methanol poisoning in the rat. *J. Toxicol. Environ. Health* **2**, 1201–1209.

Mannering, G.J. and Parks, R.E. Jr. (1975) Inhibition of methanol metabolism with 3-amino-1,2,4-triazole. *Science* **126**, 1241–1242.

McMartin, K.E., Makar, A.B., Martin, A.G., Palese, M., and Tephly, T.R. (1975) Methanol poisoning. I. The role of formic acid in the development of metabolic acidosis in the monkey and the reversal by 4-methylpyrazole. *Biochem. Med.* **13**, 319–333.

McMartin, K.E., Martin-Amat, G., Makar, A.B., and Tephly, T.R. (1977) Methanol poisoning. V. Role of formate metabolism in the monkey. *J. Pharamacol. Exp. Ther.* **201**, 564–572.

Noker, P.E. and Tephly, T.R. (1980) The role of folates in methanol toxicity. *Adv. Exp. Med. Biol.* **132**, 305–315.

Osterloh, J.D., D'Alessandro, A., Chuwers, P., Mogadeddi, H., and Kelly, T.J. (1996) Serum concentrations of methanol after inhalation at 200 ppm. *J. Occup. Environ. Med.* **38**, 571–576.

Perkins, R.A., Ward, K.W., and Pollack, G.M. (1995) A pharmacokinetic model of inhaled methanol in humans and comparison to methanol disposition in mice and rats. *Environ. Health Perspect.* **103**, 726–733.

Pirola, R.C. (1978) *Drug Metabolism and Alcohol.* University Park Press, Baltimore, MD.

Rogers, J.M., Mole, M.L., Chernoff, N., Barbee, B.D., Turner, C.I., Logsdon, T.R., and Kavlock, R.J. (1993) The developmental toxicity of inhaled methanol in the CD-1 mouse, with quantitative dose-response modeling for estimation of benchmark doses. *Teratology* **47**, 75–188.

Sedivec, V., Mraz, M., and Flek, J. (1981) Biological monitoring of persons exposed to methanol vapours. *Int. Arch. Occup. Environ. Health* **48**, 257–271.

Starr, T.B. and Festa, J.L. (2003) A proposed inhalation Reference Concentration (RfC) for methanol. *Reg. Toxicol. Pharmacol.* **38**, 224–231.

Sweeney, L.M., Kirman, C.R., Morgott, D.A., and Gargas, M.L. (2004) Estimation of interindividual variation in oxidative metabolism of dichloromethane in human volunteers. *Toxicol. Lett.* **154** (3), 201–216.

Tephly, T.R., Makar, A.B., McMartin, K.E., Hayreh, S.S., and Martin-Amat, G. (1979) Methanol – its metabolism and toxicity. *Biochem. Pharmacol. Methanol* **1**, 145–164.

USEPA (2012) Vocabulary Catalog List Detail - Integrated Risk Information System (IRIS) Glossary, Available at http://ofmpub.epa.gov/sor_internet/registry/termreg/searchandretrieve/glossariesandkeywordlists/search.do?details=&glossaryName=IRIS%20Glossary. Access date: 30 October 2012.

USEPA (2000) Trichloroethylene health risks–state of the science. Scott, C.S., Cogliano, V.J., eds., *Environ. Health Perspect.* **108**, (Suppl. 2), 159–363.

Ward, K.W., Blumenthal, G.M., Welsch, F., and Pollack, G.M. (1997) Development of a physiologically based pharmacokinetic model to describe the disposition of methanol in pregnant rats and mice. *Toxicol. Appl. Pharmacol.* **145** (2), 311–322.

Yokley, K., Tran, H.T., Pekari, K., Rappaport, S., Riihimaki, V., Rothman, N., Waidyanatha, S., and Schlosser, P.M. (2006) Physiologically-based pharmacokinetic modeling of benzene in humans: a Bayesian approach. *Risk Anal.* **26** (4), 925–943.

7 Oxidative Stress and Species Differences in the Metabolism, Developmental Toxicity, and Carcinogenic Potential of Methanol and Ethanol

PETER G. WELLS,[1,2] GORDON P. MCCALLUM,[1,†] LUTFIYA MILLER,[2,†] MICHELLE SIU,[1,†] and J. NICOLE SWEETING,[1,†]

[1]Division of Biomolecular Sciences, Faculty of Pharmacy, University of Toronto, Toronto, Ontario, Canada

[2]Department of Pharmacology and Toxicology, Faculty of Medicine, University of Toronto, Toronto, Ontario, Canada

7.1 INTRODUCTION

7.1.1 Preamble

7.1.1.1 The Regulatory Problem Methanol (MeOH) appears to be metabolized quite differently in rodents such as mice and rats than it is in humans. Nevertheless, most studies to estimate the human safety of MeOH in the developing embryo and fetus (hereafter collectively referred to as the *embryo*) have been carried out in rodents, raising the question of the accuracy, if not the relevance, of regulations based solely upon such studies.

7.1.1.2 Fundamental Question From the regulatory viewpoint, the fundamental question is whether the mechanism by which MeOH harms

† These authors contributed equally.

The Toxicology of Methanol, First Edition. Edited by John J. Clary.

the embryo is similar in rodents and humans. Since the mechanism of MeOH toxicity in the embryo remains to be established, it is not clear to what extent the reported differences between rodents and humans in the metabolism of MeOH determine its relative toxicity in the embryo of these two species. It is similarly unclear whether the molecular effects of MeOH in rodent embryos reflect those in the developing human.

7.1.1.3 Research Objectives Our objectives were to determine

(1) whether reactive oxygen species (ROS) play a role in the embryonic toxicity of MeOH in mice;
(2) whether the mechanism of this embryotoxicity is the same in the mouse as it is in a nonrodent species that, unlike the mouse, metabolizes MeOH like humans;
(3) whether MeOH oxidatively damages DNA in potential cancer target tissues of adult male mice and adults from nonrodent species that metabolize MeOH more like humans.

7.1.1.4 Approach Our approach was based on methods and models that we have previously used to determine the role of ROS in adverse embryonic effects that either arise spontaneously or are caused by drugs (Wells et al., 2009b). The mouse was used as the basic animal model and compared to a second nonrodent species that, unlike the mouse, appears to metabolize MeOH like humans, primarily via alcohol dehydrogenase (ADH1) and cytochrome P450 2E1 (CYP2E1), rather than via catalase as in rodents (Tephly et al., 1964; Makar et al., 1968; Cederbaum and Qureshi, 1982). The toxicological relevance of catalase in rodents is potentially complex because in addition to its role in MeOH metabolism, catalase also plays an antioxidative role in the detoxification of ROS (Wells et al., 2009b) (Figure 7.1). Mice were chosen as a species for which (1) MeOH teratogenicity has been well characterized (Rogers and Mole, 1997; Harris et al., 2003; Degitz et al., 2004) and (2) genetically modified strains are available for testing the contribution of specific biochemical pathways to the mechanism of teratogenesis. Rabbits were the first nonrodent species investigated as potentially

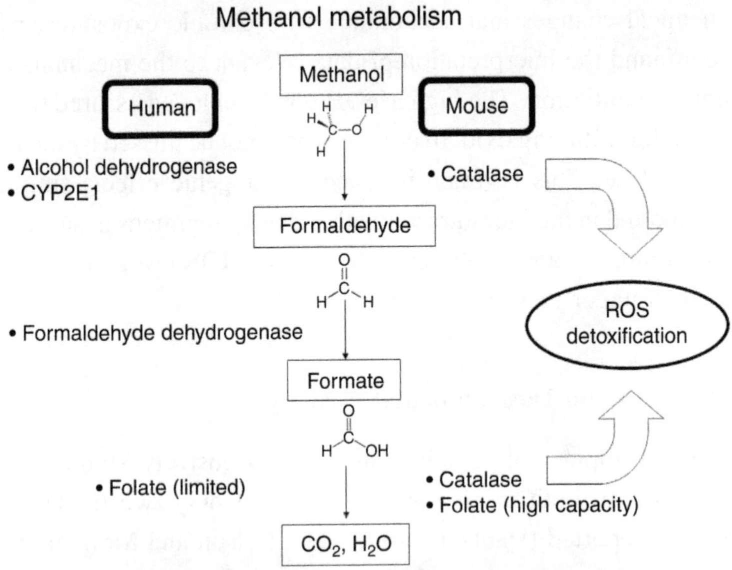

FIGURE 7.1 Species differences in the enzymes catalyzing the metabolism of methanol (MeOH) to formaldehyde and formic acid (FA) in mice and primates, including humans. In addition to its role in MeOH and FA metabolism in mice, catalase is used by all species in the detoxification of reactive oxygen species (ROS). *Source*: From Sweeting et al. (2010).

reflective of humans, based in part on their remarkable species difference from mice in susceptibility to birth defects caused by the ROS-initiating sedative drug thalidomide (Parman et al., 1999).

The intraperitoneal (i.p.) route, as opposed to inhalation or oral intake, was used in our studies to ensure that the dose delivered was as consistent as possible, in part for a precise determination of pharmacokinetic parameters, particularly across species such as mice, rabbits, and primates. Equally important, the resulting consistency in embryonic exposure allows a more reliable interpretation of results from mechanistic studies relating molecular changes to teratological outcomes. The high dose of MeOH, well above the lethal dose in humans, was based on published studies of MeOH teratogenicity in mice (Rogers et al., 2004) and retained in rabbits and primates to ensure that any teratogenic potential was not missed owing to an insufficiently high dose as well as to facilitate cross-species comparisons. MeOH was administered acutely rather than chronically to avoid the complicating secondary biochemical changes that accumulate with chronic exposures, which can confound the interpretation of data relevant to the mechanisms of teratogenic initiation. The high MeOH dose employed assured that any potential for initiating oxidative stress would not be missed by the use of an acute dose. This regimen produced teratogenic effects similar to those reported in the literature with other dosing regimens in susceptible mouse strains, so any teratologically relevant ROS involvement would be evident under these conditions.

7.1.2 Methanol Developmental Toxicity

MeOH developmental toxicity has been extensively studied in the mouse and rat, both *in vivo* and *in vitro*, but only two human cases have been reported (Hantson et al., 1997; Belson and Morgan, 2004). The underlying mechanism of this developmental toxicity in rodents is not well understood. Studies in animal models have shown different species- and strain-dependent outcomes from MeOH *in utero* exposure. The CD-1 and C57BL/6 strains of mice, and Sprague-Dawley and Long-Evans rats have all been shown to be susceptible to MeOH teratogenicity in a dose-dependent manner, but to varying degrees and with a spectrum of outcomes depending on the strain or species. However, one strain of rat, the Holtzman rat, was reported to be resistant to MeOH teratogenicity at doses of up to 3.2 g/kg bw/day with the specific dosing schedule employed in that study (Cummings, 1993) (Table 7.1). Additionally, rats appear to be less sensitive to the developmental effects of MeOH than mice, where concentrations at least fourfold higher were required in embryo culture to produce the same anomalies in rats as seen with mice (Andrews et al., 1993; Harris et al., 2003; Hansen et al., 2005). Given that the metabolism of MeOH differs between rodents and humans, it would be beneficial to determine if there is an alternative animal model better suited to predicting the risk of developmental outcomes of MeOH exposure in humans.

Teratogenicity of the related alcohol, ethanol (EtOH), is much better understood. In humans, exposure of the embryo to EtOH during gestation is known to lead to a variable spectrum of birth defects termed Fetal Alcohol Spectrum Disorders (FASD), which includes Fetal Alcohol

TABLE 7.1 Species- and Strain-Specific Malformations Following *in utero* MeOH Exposure in Mice, Rats, and Non-Human Primates

Model	Strain	ROA	MeOH Dose	Dosing Period	Assessment Parameter(s)	Outcome	Citation
Mouse	CD-1	Inhalation	1,000, 2,000, 5,000, 7,500, 10,000, 15,000 ppm, 7 h/day	GD 6–15	*GD 17*: gross morphological abnormalities and internal soft tissue anomalies	↑ exencephaly and cleft palate at ≥5,000 ppm, ↑ embryo/fetal death at >7,500 ppm, ↓ fetal weight at >10,000 ppm Dose-related ↑ in cervical rib ossification sites >2,000 ppm Established NOAEL for study at 1,000 ppm	Rogers et al. (1993) and NTP (2004)
Mouse	CD-1	Inhalation	10,000 ppm, 6 h/day	GD 6–15, 7–9, 9–11	Maternal effects, external embryonic malformations	↓ maternal body weight *GD 6–15*: ↑ resorptions, NTDs, cleft palates, and NS ↑ digit defects *GD 7–9*: ↑ resorptions, NS ↑ NTDs, cleft palates *GD 9–11*: ↑ cleft palates, NS ↑ digit defects	Bolon et al. (1993) and NTP (2004)
Mouse	CD-1	Inhalation	15,000 ppm, 6 h/day	GD 7–9	*GD 17*: cephalic NTDs, skeletal malformations	Neurological distress in 15% of dams (ataxia, circling, tilted heads, etc.) ↑ cephalic NTDs, (mostly exencephaly, some cases of anencephaly, encephalocele, holoprosencephaly, microcephaly)	Bolon et al. (1994)

(continued)

TABLE 7.1 (*Continued*)

Model	Strain	ROA	MeOH Dose	Dosing Period	Assessment Parameter(s)	Outcome	Citation
Mouse	CD-1	Inhalation	10,000 ppm, 7 h/day for single or two-consecutive-day exposure (7 h/day both days)	GD 6–13 (plug = GD 0)	*GD 17:* maternal blood profile (28 h post-inhalation period), cleft palate, cephalic NTDs	No visible intoxication of dams. Maternal plasma peak 4 mg/ml (~120 mmol/l). *Cleft Palate*: 2-day exposure, highest incidence between GD 6–7, single day highest incidence GD 7 *NTDs*: 2-day exposure highest incidence between GD 6–7, 8–9, single-day exposure highest incidence on GD 7	Rogers and Mole (1997) and NTP (2002)
Mouse	C57BL/6	i.p.	(low) 3.4 or (high) 4.9 g/kg bw total dose, split into two doses, 4 h apart	GD 7 (plug = GD 0)	*GD 17:* ophthalmic, craniofacial abnormalities	No maternal deaths or toxicity. *High dose*: ↑↑ microphthalmia, anophthalmia. ↑↑ fetuses exhibiting at least one craniofacial malformation (e.g., facial clefts). *Low dose*: ↑ microphthalmia, anophthalmia, ↑ craniofacial malformations	Rogers et al. (2004)
Rat	Sprague-Dawley	Inhalation	5,000, 10,000, 20,000 ppm, 7 h/day. (10 000, 16,000, 20,000 ppm, EtOH, 7 h/day)	GD 1–19	*GD 20*: gross morphological anomalies, skeletal abnormalities	20,000 ppm (and 10,000 ppm, to lesser extent) produced ↑ malformations (e.g., rudimentary ribs, cardiovascular defects). NS ↑ in malformations with EtOH exposure, but >maternal toxicity	Nelson et al. (1985) and NTP (2004)

Rat	Sprague-Dawley	Inhalation	200, 1,000, 5,000 ppm, average 22.7 h/day	GD 7–17	Maternal endpoints, *GD 20*: gross morphological anomalies, skeletal abnormalities	Maternal endpoints: ↑ in gestation period, ↓ post-implantation embryo survival, # live pups/litter ↑ late resorptions, ↓ litter size, ↓ fetal weight and postnatal survival rate, ↑ incidence of malformations (e.g., cardiovascular and rib defects), delayed ossification seen at ≥5,000 ppm	NEDO (1987) and NTP (2004)
Rat	Holtzman	Drinking water	1.6, 2.4, or 3.2 g/kg bw/day	GD 1–8	*GD 9*: maternal hormone levels and implantations sites *GD 11*: embryo viability, growth, and development *GD 20*: fetal weight, viability, and gross external malformations	No significant embryonic or fetal effects with MeOH treatment	Cummings (1993) and NTP (2004)
Rat	Wistar	Gavage	Well-nourished (WN) vs malnourished (MN) pregnant rats, given 2.5 g/kg bw	GD 6–15	*GD 21*: morphological and skeletal malformations	↑ skeletal malformations (e.g., extra ribs), WN control = 5.6%, WN MeOH = 45.4%, MN control = 3.8%, and MN MeOH = 38.8%, no parameters altered by MN	De-Carvalho et al. (1994)
Rat	Long-Evans	Gavage	0.0, 1.3, 2.6, or 5.2 ml/kg (1023, 2045, or 4090 mg/kg bw/day)	GD 10	*GD 20*: gross morphological anomalies (internal and external)	Dose-dependent ↑ undescended testes, exophthalmia, anophthalmia	Youssef et al. (1997) and NTP (2004)
Non-human primate	*Macaca fascicularis*	Inhalation	10, 100, 1,000 ppm, 22 h/day	Up to 30 months	Gross morphological anomalies	No teratogenic effect	NEDO (1987)

ROA, route of administration; GD, gestational day (if no other indication, plug = GD 1); NOAEL, no observable adverse effect level; NTD, neural tube defect; i.p., intraperitoneal; EtOH, ethanol; NS, nonsignificant; WN, well-nourished; MN, malnourished.

Syndrome (FAS). There are specific craniofacial abnormalities that are commonly associated with FAS, including a smooth philtrum (groove between nose and upper lip), thin upper lip, and small palpebral fissures, where the eye width is shortened (Jones and Smith, 1973). Neurological and functional impairments can also result from fetal EtOH exposure (Jones, 2011). It is difficult to determine the exact dose of EtOH in humans that leads to FASD, precluding a direct comparison of drug potency with MeOH, for which no human developmental effects have been reported. EtOH, like MeOH, is teratogenic in mice and similarly exhibits strain differences in teratogenicity (Abel, 1982; Gilliam et al., 1987; Becker and Randall, 1989; Mattson et al., 2001). Our studies further investigating species- and strain-specific differences in susceptibility to both EtOH and MeOH are discussed later in Section 7.3.

7.1.3 Carcinogenic Potential

EtOH consumed in alcoholic beverages is regarded as carcinogenic in humans based on epidemiological evidence (NTP, 2002; IARC, 2006; AICR, 2007). Mechanistic evidence supporting a causal role of acetaldehyde in alcohol-related esophageal cancer comes from studies in aldehyde dehydrogenase-2 (ALDH2)-deficient individuals who consume large quantities of alcoholic beverages (Baan et al., 2007; Brooks et al., 2009). However, there is a fundamental difference between EtOH and MeOH in that the majority of primary exposures to MeOH of toxicological interest, other than accidental poisonings from oral ingestion, will occur via exposure to vapors or dermal contact of the skin. It is important to keep this fundamental difference in mind as the exposure to EtOH in subsets of the human population will be orders of magnitude higher than that following environmental and/or industrial exposure to MeOH, and this factor is central to any assessment of risk even if these agents share a common potential mechanism of initiation owing to structural or biochemical similarities. Among other unappreciated possibilities, MeOH could theoretically cause cancer either by the covalent binding of its reactive formaldehyde metabolite to DNA, or via the formation of highly reactive and potentially toxic forms of oxygen termed reactive oxygen species (ROS) and ROS-initiated changes in signal transduction or oxidative damage to DNA. Although several studies in mice and rats have found no evidence for MeOH carcinogenesis (NEDO, 1985, 1987), one study reported an increase in lymphohematoreticular neoplasms in Sprague-Dawley rats exposed to MeOH in the drinking water (Soffritti et al., 2007). The latter study was confounded by unusual experimental conditions whereby the rat colony had a high potential for *Mycoplasma pulmonis* lung infection, was not specifically pathogen free, and had a high background level of lymphoma in the control animals (Soffritti et al., 2007; Cruzan, 2009). Our studies on carcinogenic potential are discussed later in Section 7.4 on Oxidative Stress.

7.1.4 Oxidative Stress and Other Potential Mechanisms of Toxicity

The mechanism by which MeOH causes toxicity in the developing embryo is not clearly understood, but one contributing factor may be enhanced "oxidative stress," or increased formation of ROS. These ROS include highly toxic "free radical" intermediates such as hydroxyl radicals that have been implicated in a number of human diseases and drug toxicities including cancer, neurodegenerative diseases, and birth defects (Halliwell and Gutteridge, 2007; Klaassen, 2008) (see Section 7.4).

Using a number of models involving genetically altered animals and drugs that enhance oxidative stress, results from our laboratory and others have shown that ROS can adversely affect embryonic development in at least two ways (Wells et al., 1997, 2009a, 2009b). The first is by altering the level of signals, termed "signal transduction," within embryonic cells that are involved in the activation of genes, leading to the production or suppression of proteins ultimately necessary for normal development. The second effect of ROS is to damage cellular macromolecules such as RNA, DNA, proteins, and lipid membranes, resulting in their inability to perform their normal developmental role.

Other potential mechanisms in the developmental toxicity of MeOH would include covalent binding to cellular macromolecules (RNA, DNA, proteins) by the reactive formaldehyde metabolite, which is substantially more embryotoxic in embryo culture than MeOH itself or its stable formate metabolite (Hansen et al., 2005). Additionally, as yet undetermined effects of MeOH itself have been postulated to initiate teratogenesis (NTP, 2002). Although formaldehyde remains a candidate for the proximate toxic chemical species due to its high embryotoxic potency in embryo culture (Hansen et al., 2005), the physiological relevance of treating embryos with this reactive metabolite, as distinct from having it produced metabolically within the embryo, remains unclear. However, the lower levels of embryonic formaldehyde dehydrogenase (ADH3) levels in mice compared to rats (Harris et al., 2003) may explain the greater susceptibility of mice to MeOH teratogenesis, consistent with a role for formaldehyde as the proximate chemical teratogen. The acute ocular toxicity and lethality observed in humans poisoned with MeOH appear to be due to metabolic acidosis initiated by high concentrations of the FA metabolite, which appears not to play a role in the developmental toxicity of MeOH (see Section 7.3.1).

Our studies on the role of ROS in MeOH developmental toxicity are discussed later in Section 7.4.

7.1.5 Factors Affecting the Human Relevance of Animal Models

7.1.5.1 Species Differences in Metabolism Primates, including humans, use alcohol dehydrogenase (ADH1) to oxidize MeOH to formaldehyde, whereas rodents use the peroxidative activity of catalase (Tephly et al., 1964; Makar et al., 1968; Cederbaum and Qureshi, 1982) (Figure 7.1). Both humans and rodents convert formaldehyde into formic acid (FA) by ADH3 (Teng et al., 2001; Harris et al., 2004). FA is subsequently metabolized into carbon dioxide and water by a folate-dependent dehydrogenase (Johlin et al., 1987). Folate, however, is limited in humans, leading to an accumulation of the toxic FA metabolite following exposure to high levels of MeOH (Perkins et al., 1995). Conversely, folate is not limited in rodents (Black et al., 1985), and a combined use of the peroxidative activity of catalase and the aforementioned folate-dependent pathway prevents the accumulation of the acutely toxic FA (Cook et al., 2001; Clary, 2003; Harris et al., 2004). Accordingly, the metabolism of MeOH in rodents does not reflect that in humans.

Despite these differences in metabolic pathways, the rate of MeOH metabolism appears to be similar between primates and rats following a saturating dose of MeOH (Tephly et al., 1964; Makar et al., 1968; Kavet and Nauss, 1990). This is in contrast to mice, in which the reported rate of MeOH metabolism is approximately twice as fast as that in both primates and rats (Mannering et al., 1969) (Table 7.2).

7.1.5.2 Dose of Methanol and Route of Exposure Following an 8-hour inhalation exposure of 5000 ppm MeOH, the blood MeOH concentration and daily dose have been determined for mice and rodents and projected for humans, where mice exhibit a 17-fold higher calculated plasma concentration than humans (Perkins et al., 1995). Different routes of MeOH exposure in various animal models and humans via i.p. injection, oral ingestion, and inhalation may produce different peak plasma concentrations of MeOH, not to mention its metabolites, in both the mother and developing embryo, which may alter developmental risks.

7.2 SPECIES DIFFERENCES IN METHANOL METABOLISM

7.2.1 Enzymes and Pathways

Additional comments on species differences particularly in comparison to humans are provided in Section 7.1.5.1.

7.2.1.1 Alcohol Dehydrogenase (ADH1) ADH1 (EC 1.1.1.1) is an oxidoreductase enzyme responsible for the metabolism of alcohols into their corresponding aldehydes, using nicotinamide adenine dinucleotide (NAD^+) as a cofactor (Lutwak-Mann, 1938) (Figure 7.2). The enzyme is constructed of two subunits, each subunit being one of six possible subunits (α, β, γ, π, χ, or σ), where the class 1 isoenzymes of ADH1 are of the $\alpha\alpha$, $\alpha\beta$, or $\alpha\gamma$ variety (Kedishvili et al., 1995). ADH1 is highly expressed in the liver, and each isoenzyme has a varying capacity to metabolize alcohols (Wagner et al., 1983).

TABLE 7.2 Pharmacokinetic Data for Methanol (MeOH) Metabolism in Rodents, Humans, and Non-human Primates

			Peak Serum/Plasma Concentration					
Model	MeOH Dose	ROA	MeOH	FA	MeOH Clearance	FA AUC	Half-Life	Citation
Mouse	2.0 g/kg bw	i.p.	79 mM	0.31 mM	1.53 ml/min × kg bw	3 mM × h	N/D	Sweeting et al. (2010)
Rat	100 mg/kg bw	i.v.	3.4 mM	0.08 mM	1.2 ml/min	3 mM × h	1.4 h	Horton et al. (1992)
Rabbit	2.0 g/kg bw	i.p.	114 mM	0.28 mM	0.57 ml/min × kg bw	10.63 mM × h	N/D	Sweeting et al. (2010)
Human	200 ppm over 4 h (equivalent to 4 mg/kg)	Inhalation	0.2 mM at the end of 4-h exposure period	0.31 mM at the end of 4-h exposure period	1.85 ml/min × kg bw	N/D	Nonsaturating dose: 2–4 h Saturating dose: 50 h	Haffner et al. (1992), Osterloh et al. (1996), and Hovda et al. (2007)
Primate	2.0 g/kg bw	i.p.	95 mM at 6 h post injection	2 mM at 6 h post injection	N/D	8 mM × h at 6 h post injection	3 h (2000 ppm inhalation)	Horton et al. (1992) and Sweeting et al. (2010)

ROA, route of administration; AUC, area under the plasma concentration–time curve; i.p., intraperitoneal; i.v., intravenous; MeOH, methanol; FA, formic acid; N/D, not determined.

$$CH_3OH + NAD^+ \xleftrightarrow{ADH} CH_2O + NADH + H^+$$

FIGURE 7.2 Role of alcohol dehydrogenase (ADH1) in the metabolism of MeOH. MeOH (CH_3OH) is oxidized in humans and primates by ADH1 into formaldehyde (CH_2O) using NAD^+ as a cofactor.

The physical properties of ADH1 differ considerably among species, where the affinity of MeOH for ADH1 varies greatly (Mani et al., 1970; Pietruszko, 1975a,b) (Table 7.3). In humans and non-human primates, MeOH is preferentially oxidized to formaldehyde using the ADH1 $\beta_1\beta_1$ isoenzyme (Mannering et al., 1969; Wagner et al., 1983; Blomstrand and Ingemansson, 1984). Saturable elimination results from a limited source of the ADH1 cofactor NAD^+ and occurs following doses greater than 0.1 g/kg MeOH in humans and non-human primates (Horton et al., 1992; Rang et al., 2003). In contrast, rodent ADH1 makes a minimal contribution to MeOH metabolism, as demonstrated in ADH1-deficient mice (Bradford et al., 1993). The underlying mechanism responsible for the inability of rodents to efficiently metabolize MeOH by ADH1 remains unknown; however, species differences in the ADH1 gene or protein expression (Buhler et al., 1984) and differing isoenzymes (Algar et al., 1983) have been implicated.

ADH1 activity using EtOH as a substrate has been well characterized in humans and rodents; however, the use of MeOH as a substrate is less understood, and therefore, assessment of species differences in ADH1 activity generally refers to the former substrate (Pikkarainen and Raiha, 1967; Algar et al., 1983; Harris et al., 2003). An interspecies comparison of ADH1 activity with EtOH demonstrates parallels between humans and rodents in the increasing ADH1 activity throughout gestation and postnatally, with highest levels being reached in

TABLE 7.3 Comparison of Enzyme Kinetic Parameters for Alcohol Dehydrogenase (ADH1) with MeOH As a Substrate in Mice, Rats, and Humans

Model	V_{max}	K_m	Citation
Mouse	0.06 mmol/min/kg bw	48.7 mM	Ward et al. (1995)
Rat	0.03 mmol/min/kg bw	48.7 mM	Ward et al. (1995)
Human	0.6 Units/μmol of active site	6 mM	Wagner et al. (1983)

V_{max}, maximum rate of enzyme reaction; K_m, substrate concentration at $\frac{1}{2}$ V_{max}.

TABLE 7.4 Changes in ADH1 Activity Throughout Development in Mice, Rats, and Humans

Model	Age	Organ	Activity of ADH1	Citation
Mouse	7–12 somites	Whole embryo	0.1–0.3 nmol/min/mg protein	Harris et al. (2003)
	20–22 somites	Whole embryo	0.5 nmol/min/mg protein	Harris et al. (2003)
	Adult	Liver	1360 nmol/min/mg protein	Algar et al. (1983)
Rat	7–12 somites	Whole embryo	0.25–0.6 nmol/min/mg protein	Harris et al. (2003)
	20–22 somites	Whole embryo	0.49–0.62 nmol/min/mg protein	Harris et al. (2003)
Human	Fetus 2–6 months	Liver	149 mU/g liver wet weight	Pikkarainen and Raiha (1967)
	Infant 0.3–7 months	Liver	579 mU/g liver wet weight	Pikkarainen and Raiha (1967)
	Children and adults 2–50 years	Liver	2410 mU/g liver wet weight	Pikkarainen and Raiha (1967)

adulthood. However, human ADH1 activity is consistently higher than that found in both mice and rats (Table 7.4).

7.2.1.2 Catalase Catalase (EC 1.11.1.6) is a tetramer containing four strongly bound molecules of nicotinamide adenine dinucleotide phosphate (NADPH) (Kirkman and Gaetani, 1984). Each tetrameric unit contains four heme groups, which determine the function of the enzyme on the basis of their redox state. Consequently, catalase can exist in three states: (1) the resting state (containing ferric heme groups), which is responsible for the antioxidative activity of catalase; (2) Compound I (containing Fe^{4+} heme groups), which is responsible for the peroxidative activity of catalase; and (3) Compound II, which is the inactive form of the enzyme (Goyal and Basak, 2010). The tightly bound NADPH molecules prevent the formation of Compound II (Kirkman and Gaetani, 1984).

$$\text{(a)}\quad CH_3OH + H_2O_2 \xrightarrow{\text{Catalase}} CH_2O + 2H_2O$$

$$\text{(b)}\quad HCOOH + H_2O_2 \xrightarrow{\text{Catalase}} CO_2 + 2H_2O$$

FIGURE 7.3 Role of catalase in the metabolism of MeOH and FA: peroxidative activity. Using hydrogen peroxide (H_2O_2) as a cofactor, rodent catalase peroxidatively oxidizes (a) MeOH (CH_3OH) into formaldehyde (CH_2O) and (b) FA (HCOOH) into carbon dioxide and water.

Peroxidative Role In the presence of hydrogen peroxide (H_2O_2) and a hydrogen donor such as an alcohol, the Compound I state of catalase oxidizes the alcohol to its aldehyde, in the process reducing catalase back to its resting state (Figure 7.3). In rodents, MeOH is metabolized by the peroxidative activity of catalase as opposed to ADH1, forming the toxic formaldehyde metabolite (Karinje and Ogata, 1990; Bradford et al., 1993). Conversely, humans and non-human primates do not metabolize MeOH using catalase (Mannering et al., 1969), likely because of low levels of endogenous H_2O_2 (Oshino et al., 1975). In rodents, catalase is also responsible for the peroxidative oxidation of FA to carbon dioxide and water (Smith and Taylor, 1982; Von Burg, 1994) (Figure 7.3)—a process that does not occur in humans.

In rodent embryos, peroxidative catalase activity increases with gestational age, and comparatively, the mouse embryo has consistently higher specific catalase activity than the rat embryo (Harris et al., 2003) (Table 7.5). The peroxidative activity of catalase follows Michaelis–Menten kinetics, where rat catalase demonstrates an affinity for MeOH of $\sim$1.5 mM (Perkins et al., 1995).

Antioxidative Role In all species, catalase acts as an antioxidant by scavenging H_2O_2 (Figure 7.4) (Halliwell and Gutteridge, 2007; Wells et al., 2009b). Similar to its peroxidative activity, antioxidative catalase activity increases throughout gestation and into adulthood in both humans and rodents, where mouse embryonic catalase activity is approximately 4% of adult activity (Table 7.6) (Winn and Wells, 1999; Abramov and Wells, 2011). Furthermore, substantial species differences in antioxidative catalase activity have been documented (Reddy et al., 1984; Sweeting and Wells, 2010). Unlike the peroxidative

TABLE 7.5 Changes in Peroxidative Catalase Activity, Using MeOH As a Substrate, Throughout Development in Mice and Rats

Model	Age	Organ	Peroxidative Activity of Catalase	Citation
Mouse	6–12 somites	Whole embryo	2 nmol formaldehyde formed/min/mg protein	Harris et al. (2003)
	21–28 somites	Whole embryo	21 nmol formaldehyde formed/min/mg protein	Harris et al. (2003)
	Adult	Blood	32.5 U/mg protein	Ishii et al. (1996)
Rat	6–12 somites	Whole embryo	7.5 nmol formaldehyde formed/min/mg protein	Harris et al. (2003)
	21–28 somites	Whole embryo	17 nmol formaldehyde formed/min/mg protein	Harris et al. (2003)
	Adult	Blood	55 U/mg hemoglobin	Breinholt et al. (2000)

activity of catalase, antioxidative catalase activity follows Michaelis–Menten kinetics only at low concentrations of H_2O_2 because of the high reaction rate of this process and the ability of high H_2O_2 levels to inactivate the enzyme. Accordingly, only theoretical kinetic parameters for antioxidative catalase may be calculated, where an apparent K_m of 25 mM has been reported for H_2O_2 (Vetrano et al., 2005). As such, the apparent affinity of rodent catalase for H_2O_2 is approximately 20-fold lower than that for MeOH (see section titled Peroxidative Role), indicating that MeOH will be preferentially metabolized over H_2O_2. With less H_2O_2 being scavenged, the potential for ROS formation is enhanced, leading to an increased possibility of oxidative damage.

7.2.1.3 Cytochrome P450 (CYP) 2E1 Alcohols including EtOH and MeOH can be metabolized by a microsomal EtOH oxidizing system (MEOS) to their respective aldehydes (Teschke et al., 1974). The

$$2H_2O_2 \xrightarrow{\text{Catalase}} 2H_2O + O_2$$

FIGURE 7.4 Role of catalase in the metabolism of reactive oxygen species (ROS): antioxidative activity. H_2O_2 is scavenged by the antioxidative function of catalase in all species, forming the nontoxic metabolites of water and oxygen.

TABLE 7.6 Changes in Antioxidative Catalase Activity Throughout Development in Rodents, Humans, and Primates

Model	Age	Organ	Antioxidative Activity of Catalase	Citation
Mouse (CD-1)	GD 12 embryo	Whole embryo	13 U/mg protein	Winn and Wells (1999)
	Adult	Liver	400 U/mg protein	Winn and Wells (1999)
Mouse (C57BL/6)	GD 9 embryo	Whole embryo	20 U/mg protein	Abramov and Wells (2011)
	GD 13 embryo	Whole embryo	33 U/mg protein	Abramov and Wells (2011)
	GD 17–19 embryo	Brain	21 U/mg protein	Abramov and Wells (2011)
	Adult	Liver	154 U/mg protein	Abramov and Wells (2011)
Mouse (C3H)	GD 9 embryo	Whole embryo	25 U/mg protein	Abramov and Wells (2011)
	GD 13 embryo	Whole embryo	40 U/mg protein	Abramov and Wells (2011)
	GD 17–19 embryo	Brain	36 U/mg protein	Abramov and Wells (2011)
	Adult	Liver	637 U/mg protein	Abramov and Wells (2011)
Rat	6–12 somites	Whole embryo	75 nmol formaldehyde formed/min/mg protein	Harris et al. (2003)
	21–28 somites	Whole embryo	17 nmol formaldehyde formed/min/mg protein	Harris et al. (2003)
Human	Newborn	Umbilical cord blood	19 U/g hemoglobin	Saker et al. (2008)
	Adult (mother)	Blood	68 U/g hemoglobin	Saker et al. (2008)
Primate	Adult	Liver	86–105 U/mg protein	Reddy et al. (1984)

GD, gestational day.

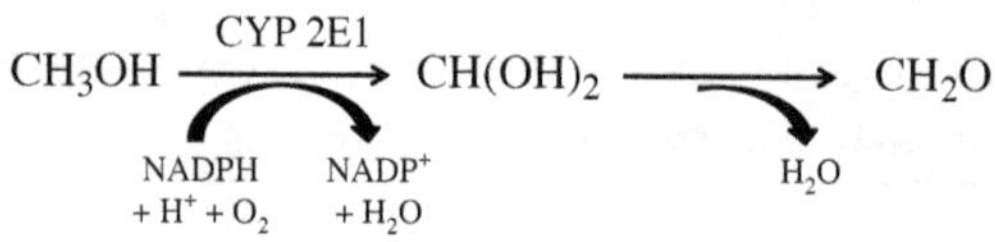

FIGURE 7.5 Role of cytochrome P450 2E1 (CYP2E1) in the metabolism of MeOH. In the presence of NADPH and molecular oxygen, CYP2E1 oxidizes MeOH into formaldehyde (CH_2O).

MEOS system principally relies on the cytochromes P450 (CYP) isozymes CYP2E1 and CYP1A2 for alcohol metabolism and is dependent on the presence of molecular oxygen and NADPH as the cofactor (Kunitoh et al., 1993; Lieber, 2004) (Figure 7.5).

7.2.1.4 Formaldehyde Dehydrogenase (ADH3) The half-life of formaldehyde is approximately 1.5 minutes (McMartin et al., 1979) due to its rapid oxidation into FA and/or the formation of macromolecular adducts (Shaham et al., 1996). Using aldehyde dehydrogenase (ALDH2), formaldehyde can be metabolized into FA in the mitochondria (Teng et al., 2001). Additionally, ADH3 can oxidize formaldehyde into FA in both humans and rodents through a glutathione (GSH)-dependent mechanism (Uotila and Koivusalo, 1974; Harris et al., 2004) (Figure 7.6). ADH3 is approximately 100-fold more efficient at oxidizing formaldehyde than ALDH2, suggesting that it is the predominant pathway for formaldehyde metabolism *in vivo* (Teng et al., 2001).

During development, rat and mouse embryos have similar ADH3-specific activity (Harris et al., 2003).

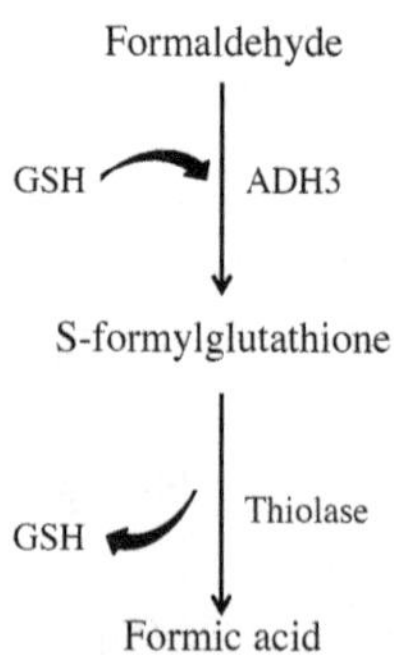

FIGURE 7.6 Role of formaldehyde dehydrogenase (ADH3) in the metabolism of formaldehyde. ADH3 conjugates glutathione (GSH) to formaldehyde forming S-formylglutathione, which is subsequently metabolized into FA through the removal of GSH by thiolase.

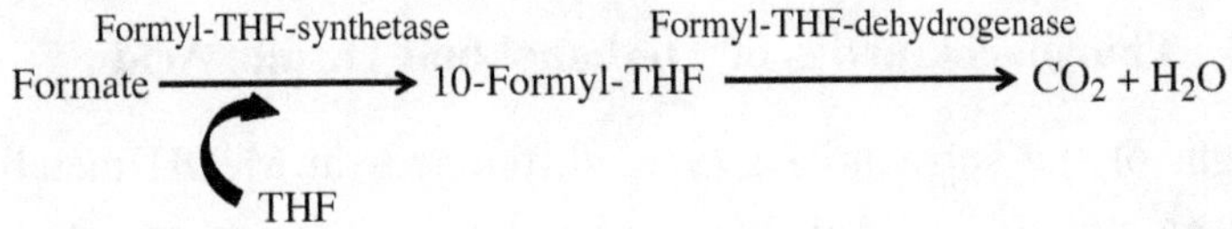

FIGURE 7.7 Role of folate in the metabolism of FA. In the presence of tetrahydrofolate (THF), formate is metabolized into 10-formyl-THF by 10-formyl-THF-synthetase. 10-Formyl-THF is subsequently converted into CO_2 and H_2O by 10-formyl-THF dehydrogenase.

7.2.1.5 Folate-dependent dehydrogenase Formic acid (FA) is metabolized to carbon dioxide and water by a folate-dependent dehydrogenase. Using 10-formyl-tetrahydrofolate (THF) synthetase, FA is converted into 10-formyl-THF, which in turn is converted into carbon dioxide and water by 10-formyl-THF dehydrogenase (Black et al., 1985) (Figure 7.7). This metabolic pathway is maintained in all species; however, varying levels of the cofactor THF limit the efficiency of FA oxidation (McMartin et al., 1981). Specifically, the mouse has over fourfold higher basal THF levels than humans, effectively decreasing the accumulation of FA and its associated toxicities (Sokoro et al., 2008) (Table 7.7). In primates, the THF-dependent metabolic pathway of FA oxidation becomes saturated with doses of MeOH exceeding 300 mg/kg (Kavet and Nauss, 1990).

In addition to the enzyme-catalyzed metabolism of FA discussed above, there are also H^+/formate co-transporters in the renal tubes, which could potentially be present in the yolk sac and placenta and subsequently affect species differences in the developmental susceptibility to FA exposure (Hovda et al., 2007).

TABLE 7.7 Species Differences in Hepatic Levels of Tetrahydrofolate (THF)

Model	THF, nmol/g	Citation
Mouse	42	Sokoro et al. (2008)
Rat	11	Sokoro et al. (2008)
Monkey	8	Sokoro et al. (2008)
Human	7	Sokoro et al. (2008)
Micropig	2	Sokoro et al. (2008)

THF, tetrahydrofolate.

7.2.2 Pharmacokinetics of Methanol and Formic Acid

In light of the substantial species differences in MeOH metabolism between humans and rodents (see Section 7.2.1 for more details), we sought to establish if human MeOH metabolism might be better reflected in rabbits than rodents. The plasma pharmacokinetics of MeOH and FA were compared in male CD-1 mice, New Zealand white (NZW) rabbits and cynomolgus monkeys over time (24, 48, and 6 hours, respectively) following a single i.p. injection of 0.5 or 2 g/kg MeOH or its saline vehicle. Following the 2 g/kg dose, MeOH exhibited zero-order elimination kinetics in all three species, with comparable peak plasma concentrations and a 2.5-fold higher clearance in mice than rabbits (Sweeting et al., 2010) (Figures 7.8 and 7.9). FA

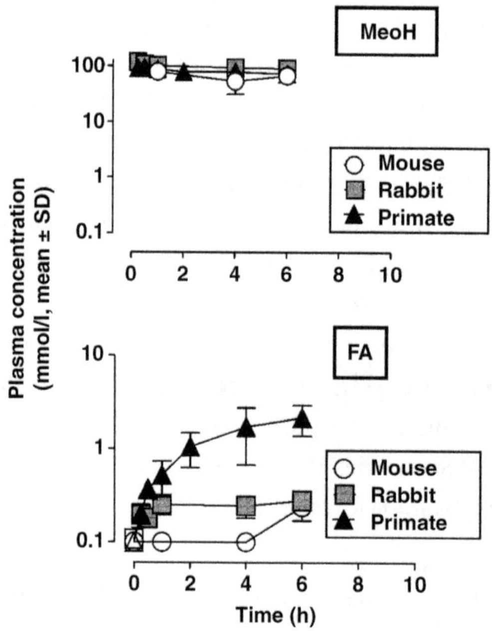

FIGURE 7.8 MeOH and FA disposition in male mice, rabbits, and primates over a 6-hour time period, following a single dose of 2 g/kg MeOH, plotted on a semilog scale. MeOH was administered through an i.p. injection as a 20% solution in sterile saline, and plasma samples were analyzed by GC (mice and rabbits) or by spectrometry (primates) for MeOH or FA concentrations. Each time point represents the mean of three to six mice, three rabbits, or three primates for treated groups. Data points with open symbols represent samples with a MeOH or FA concentration below the level of detection and were arbitrarily assigned a value of 0.1 mmol/l. *Source*: From Sweeting et al. (2010).

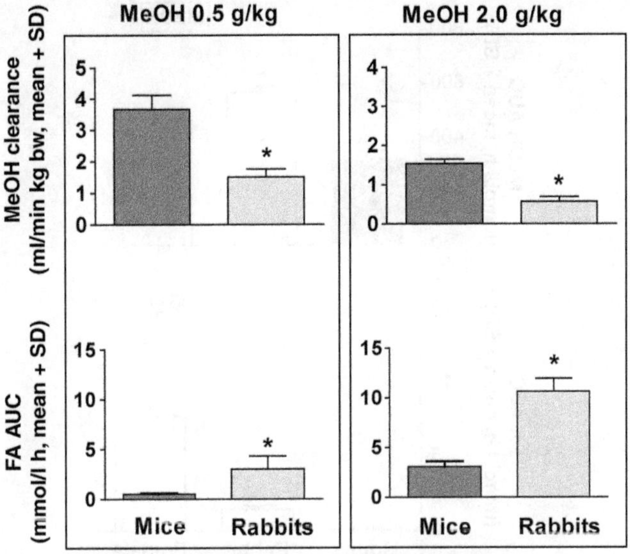

FIGURE 7.9 MeOH clearance (Cl) and FA areas under the plasma concentration–time curve (AUC) in male mice and rabbits following a single dose of 0.5 or 2 g/kg MeOH. Cl and AUC values were calculated from the MeOH and FA pharmacokinetics curves of animals dosed with 0.5 or 2 g/kg MeOH (20% solution in sterile saline) through an i.p. injection. Saline groups not shown (all not detectable). Each time point represents the mean of three to six mice or three rabbits for treated groups. The symbol $*$ indicates measures for rabbit MeOH Cl and FA AUC that were different from the respective mouse values ($p < 0.05$). *Source*: From Sweeting et al. (2010).

accumulation within 6 hours in primates was fivefold and 43-fold higher than in rabbits and mice, respectively, with accumulation being 10-fold higher in rabbits than mice (Figure 7.10). Over 48 hours, FA accumulation was nearly fivefold higher in rabbits than mice (Figure 7.9). Low-dose MeOH in mice and rabbits resulted in similarly saturated MeOH elimination in both species, but with approximately twofold higher clearance rates in mice (Figure 7.9). Furthermore, FA accumulation was 3.8-fold higher in rabbits than mice (Figure 7.9), suggesting that rabbits more closely than mice reflected primates for *in vivo* MeOH metabolism, and particularly FA accumulation (Sweeting et al., 2010).

MeOH and FA pharmacokinetics were also studied in two additional mouse strains: C57BL/6J and C3H mice. In comparison to the previously mentioned CD-1 mouse study, employing the same acute dosing

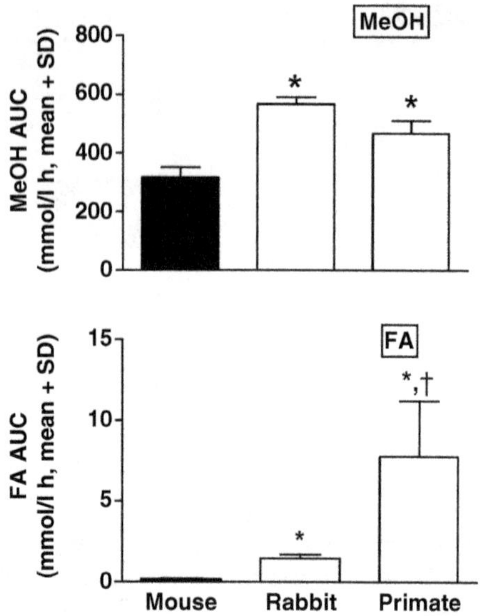

FIGURE 7.10 Six-hour MeOH and FA AUC in male mice, rabbits, and primates following a single dose of 2 g/kg MeOH. A total of three to six mice, three rabbits, and three primates were sampled at each time point for treated groups. The symbol * indicates that the 6-hour MeOH AUC and FA AUC were different from the respective mouse values. The symbol † indicates that the 6-hour FA AUC in primates was different from that in rabbits ($p < 0.05$). *Source*: From Sweeting et al. (2010).

regimen, C57BL/6J and C3H mice both exhibited pharmacokinetic profiles similar to that in the CD-1 strain (Figure 7.11) (Sweeting et al., 2011). Peak plasma concentrations and AUC levels were identical between mouse strains over the initial 12-hour period of sample collection.

Our results can be compared with other pharmacokinetic studies performed in other species and through different routes of administration (Table 7.2). MeOH and FA peak concentrations in our mice administered (i.p.) 20 times the dose given to rats (i.v.) were approximately 23-fold and fourfold higher, respectively, in the mice (Horton et al., 1992). However, the MeOH clearance rates and FA AUC values found in our mouse studies (2 g/kg MeOH i.p.) were comparable to those found in the rat (100 mg/kg MeOH i.v.). MeOH plasma concentration levels in humans at which toxicity is observed have not been

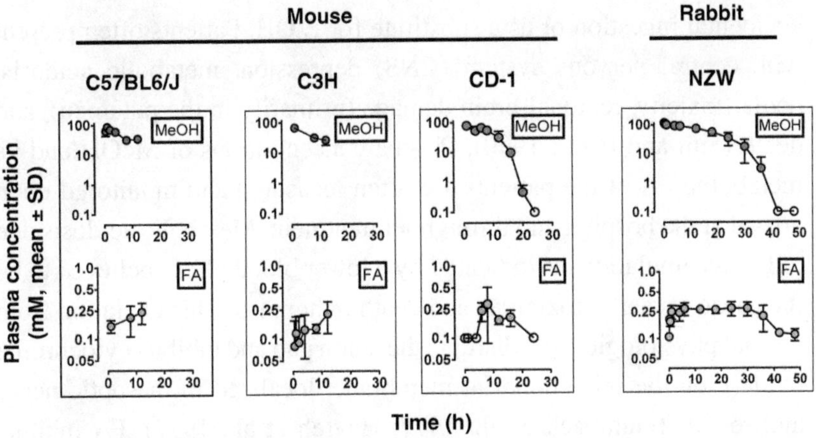

FIGURE 7.11 MeOH and FA pharmacokinetics in male C57BL/6J, C3H and CD-1 mice, and New Zealand white (NZW) rabbits. C57BL/6J and C3H mice were sampled over a 12-hour time period, CD-1 mice over 24 hours and NZW rabbits over 48 hours, following a single intraperitoneal (i.p.) dose of 2 g/kg MeOH. MeOH was administered as a 20% solution in sterile saline, and plasma samples were analyzed by GC for MeOH or FA concentrations. A total of three to six mice and three rabbits were sampled at each time point. Data points with open symbols represent samples with a MeOH or FA concentration below the level of detection and were arbitrarily assigned a value of 0.1 mM. *Source*: From Sweeting et al. (2011).

reported above 70 mM, which is below the peak concentration seen in mice in which no toxicity is observed, highlighting the species difference in susceptibility to acute toxicity. Non-human primate MeOH and FA peak plasma concentrations, as described earlier, have been reported to be higher than those observed in humans, with levels up to 90 and 2 mM, respectively (Horton et al., 1992; Sweeting and Wells, 2010) (Table 7.2).

7.3 SPECIES AND STRAIN DIFFERENCES IN METHANOL TOXICITY

7.3.1 Acute Metabolic Acidosis, Ocular Toxicity, and Death

In humans, acute rather than chronic or developmental toxicities of MeOH are predominantly reported in the literature, often due to accidental ingestion or as a substitute for EtOH. Patients often present with central nervous system (CNS) depression, metabolic acidosis, ocular toxicity, regional brain damage (primarily in the putamen), and death (Suit and Estes, 1990). Plasma concentrations of MeOH and its metabolite FA of the patients are often measured and monitored upon arrival at the hospital and throughout treatment. Metabolic acidosis due to FA accumulation, as indicated by a lower blood pH, is believed to be the cause of ocular toxicities and death in humans. This deviation from normal physiological pH disrupts the anion gap and inhibits cytochrome oxidase on the mitochondrial membrane, localized in the optic nerve and retina (Baumbach et al., 1977; Hayreh et al., 1977). FA inhibits cytochrome oxidase when at concentrations between 5 and 30 mM, which overlaps with concentrations observed in patients demonstrating acute MeOH intoxication (Nicholls, 1975, 1976). If cells are sufficiently damaged, swelling and edema result, which can compress the optic nerves and potentially lead to cell death. As metabolic acidosis worsens, further symptoms may include shallow respiration, cyanosis, coma, and electrolyte disturbances, which eventually lead to death (Suit and Estes, 1990). The blood MeOH concentration required to cause these acute effects vary, but has been reported to fall within the range of 6–16 mM for CNS effects, 16–47 mM for ocular toxicity, and 42–62 mM for fatalities (IPCS, 1997) (Figure 7.12). These concentrations, however, are an inaccurate representation of the actual blood levels required to cause these acute effects of MeOH toxicity, as several factors must be taken into consideration: (1) the amount and composition of the ingested MeOH solution is unknown; (2) the time since initial ingestion is unknown, and therefore it is unknown how much of the MeOH has already been metabolised; and (3) patients will receive some form of intervention to alleviate the symptoms of MeOH poisoning. This includes treatment with EtOH, which acts as a competitive inhibitor of ADH1 owing to its higher affinity than MeOH for the metabolizing enzyme, ADH1 (Mani et al., 1970; Ekins et al., 1985). By saturating ADH1, further metabolism of MeOH and the subsequent acute toxicities due to FA accumulation are inhibited. Hemodialysis, treatment with bicarbonate for acidotic patients, or treatment with drugs

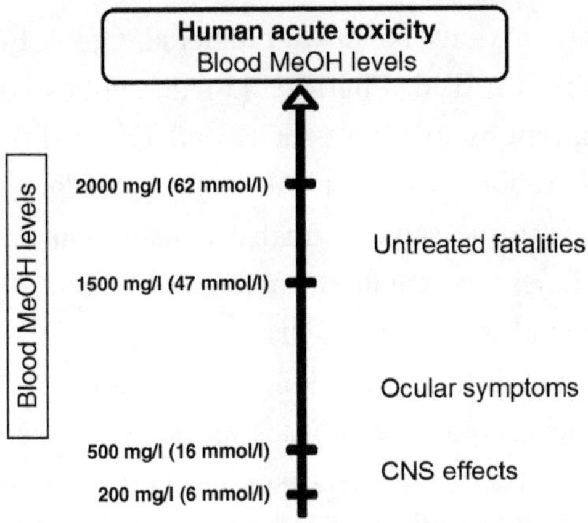

FIGURE 7.12 Blood MeOH levels measured from human cases of acute toxicity. Blood MeOH levels (mmol/l and equivalent mg/l) with corresponding symptoms of toxicity. *Source*: Modified from WHO (1997), Environmental Health Criteria: http://www.inchem.org/documents/ehc/ehc/ehc196.htm#SectionNumber:1.3 Access date: 03/10/2012.

such as fomepizole, which is a competitive inhibitor of ADH1, are also common interventions (Hovda and Jacobsen, 2008; Brent, 2009).

MeOH acute toxicity in animal models is completely different. Although primates, as expected, exhibit similar symptoms (ocular toxicity, metabolic acidosis) (Baumbach et al., 1977), rodents and rabbits do not exhibit either of these manifestations (Roe, 1955). Our laboratory has performed single dose (2 g/kg MeOH), 2-day and 15-day (daily 2 g/kg doses of MeOH on consecutive days) pharmacokinetic studies which found no accumulation of MeOH or FA in mice, whereas a sustained accumulation of both MeOH and FA was observed in rabbits following a single dose of MeOH (Sweeting and Wells, 2010; McCallum et al., 2011a,b). However, as previously reported, no evidence of ocular toxicity or metabolic acidosis was observed in either mice or rabbits, despite greater FA accumulation in the latter, and both species appeared normal throughout the studies.

Although acute MeOH toxicity in humans is thought to be caused by FA accumulation and the resulting metabolic acidosis, the mechanism of developmental toxicity is not well understood, but FA accumulation is unlikely to be involved. Pharmacokinetic studies comparing blood plasma concentrations of both MeOH and FA in different strains of mice and NZW rabbits, together with complementary *in utero* MeOH exposure studies in the same animals, indicate that pharmacokinetics alone are insufficient to explain the marked differences in susceptibility between strains and species to MeOH developmental toxicity (Sweeting et al., 2010, 2011). Similarly, among mouse strains (C57BL/6J, C3H, and CD-1), pharmacokinetic profiles are similar (see Section 7.2.2), yet each strain has a different susceptibility to MeOH *in utero* exposure; C57BL/6J mice and CD-1 mice are susceptible to morphological anomalies, and even these strains exhibit strain-specific abnormalities, whereas C3H mice are relatively resistant (Sweeting et al., 2011) (see Section 7.3.2). Only morphological abnormalities were measurably affected by MeOH administration, whereas litter parameters such as litter size or the incidence of resorptions and stillbirths were not altered. Furthermore, NZW rabbits exhibited a pharmacokinetic profile more similar to that of humans, in that FA levels accumulated and were sustained for more than twice the time observed in mice (Sweeting et al., 2011) (Figure 7.11). Rabbits also proved to be resistant to MeOH developmental toxicity, despite their substantially different metabolism of MeOH, including FA accumulation (Sweeting et al., 2011).

7.3.2 Teratogenesis

Species differences in teratogenicity are clearly exemplified in the case of thalidomide, whereby rodents are resistant while rabbits, non human primates, and humans are susceptible. Additionally, *in utero* thalidomide exposure significantly increased DNA oxidation in rabbit embryos but not in mouse embryos, suggesting oxidatively damaged DNA may be involved in the observed species difference (Parman et al., 1999).

There have been only two documented cases of human MeOH developmental toxicity. Animal models, mostly rodents, have shown species- and strain-specific sensitivity to MeOH during pregnancy, with a variable spectrum of morphological abnormalities. Rodent studies have demonstrated that the mouse is more sensitive than the rat to MeOH developmental toxicity in embryo culture, where concentrations at least fourfold higher were required to obtain the same outcomes in rats as in mice (Andrews et al., 1993; Harris et al., 2003; Hansen et al., 2005). This difference in susceptibility between the rat and mouse may potentially be explained at least in part by differences in MeOH metabolizing enzymes. Although mouse embryos express lower levels of ADH1, responsible for converting the parent drug MeOH to formaldehyde, the embryonic level of ADH3, responsible for converting formaldehyde to FA, is lower in mice than in rats, potentially resulting in higher formaldehyde levels in the mouse embryo (Harris et al., 2003). Formaldehyde when added directly to the culture medium is 1000-fold more embryotoxic than either MeOH or its FA metabolite in rat embryo culture (Hansen et al., 2005). Addition of all three chemical species to whole embryo culture decreased embryonic GSH levels, but only MeOH and formaldehyde caused dysmorphogenesis (Harris et al., 2004). Furthermore, depletion of embryonic GSH by pretreatment with L-buthionine-*S*,*R*-sulfoximine (BSO), an inhibitor of GSH synthesis, exacerbated these embryotoxic effects, suggesting a protective role for GSH-dependent detoxification of formaldehyde (Harris et al., 2004). Combined with the enhanced susceptibility of mouse embryos compared to rats, these studies implicate oxidative stress in the mechanism of MeOH teratogenesis (see Section 7.4).

A single primate study found no evidence for a teratogenic effect with inhaled MeOH at the doses employed (NEDO, 1987), but this study is yet to be replicated. Moreover, the New Energy Development Organization (NEDO) results were originally judged to be inadequately reported (NTP, 2002) although more information has since been made available through the Methanol Institute (http://www.methanol.org). Rodent models, however, have been extensively explored. Both rats and mice, as previously mentioned, are susceptible to MeOH teratogenicity, albeit to different degrees. Rats and mice exhibit different morphological outcomes, which are dependent on the rodent strain employed, the gestational day (GD) of exposure and the route of administration. The rabbit is another potential candidate as a model for human MeOH developmental toxicity. Parallel studies in C57BL/6J, C3H mice, and

NZW rabbits reveal that, similar to C3H mice, rabbits are resistant to MeOH teratogenicity despite differences in pharmacokinetics and MeOH metabolism between these species. There was no measurable effect of MeOH on any developmental parameters (litter or morphological), similar to C3H mice, with both species being exposed to similar dosing schedules during gestation (Sweeting et al., 2011) (Figures 7.13 and 7.14).

In addition to species differences in MeOH teratogenicity, there are also strain-specific malformations that result from MeOH exposure during gestation (Table 7.1). Studies primarily in rodents have outlined

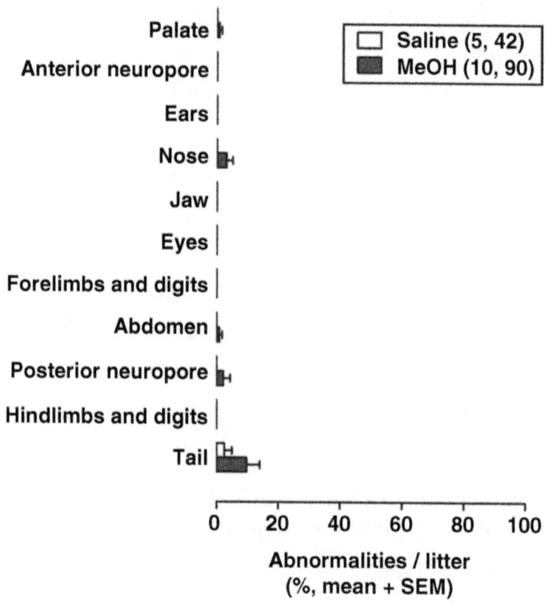

FIGURE 7.13 MeOH-initiated birth defects in NZW rabbits. Pregnant does were treated i.p. with two doses of 2 g/kg MeOH (20% solution in saline) or its saline vehicle, with an 8-hour interval, on GD 7 or 8. Does were euthanized on GD 29 for assessment of fetuses. Fetal outcomes from GD 7 and 8 MeOH exposures were not different ($p > 0.05$), and these data were combined. Fetal outcomes from GD 7 and 8 MeOH exposure were not different ($p > 0.05$), and these data were combined (x,y = number of litters, number of fetuses). *Source*: From Sweeting et al. (2011).

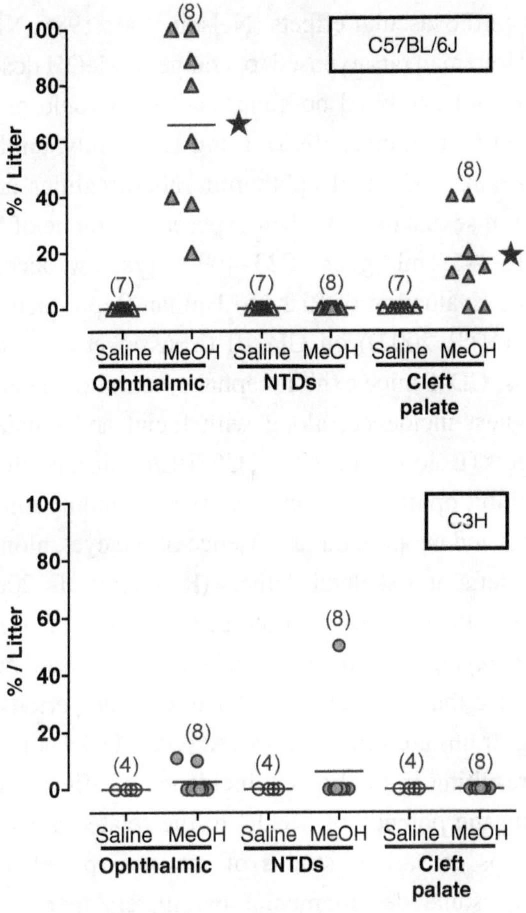

FIGURE 7.14 MeOH-initiated birth defects in C57BL/6J and C3H mice. Pregnant dams were treated i.p. with two doses of 2 g/kg MeOH (20% solution in saline) or its saline vehicle, with a 4-hour interval on 8. Dams were euthanized on GD 19 for assessment of fetuses. Each data point represents one litter, with the mean designated by a horizontal bar. The number in parentheses represents the number of litters assessed. The symbol ★ indicates a difference from the respective saline control ($p < 0.01$ for ophthalmic defects, $p < 0.05$ for cleft palates). *Source*: From Sweeting et al. (2011).

these differences, emphasizing the difficulty in identifying the most representative animal model for MeOH developmental toxicity in humans. Sprague-Dawley rats exposed to varying concentrations of inhaled MeOH (5000–20,000 ppm for 7 hours/day) exhibited only skeletal and cardiovascular effects (Nelson et al., 1985; NEDO, 1987). In contrast, Holtzman rats exposed to a range of MeOH doses by gavage (1.6–3.2 g/kg bw) exhibited no significant embryonic or fetal effects with treatment (Cummings, 1993). Long-Evans rats, unlike the aforementioned strains, exhibited ophthalmic abnormalities and abnormal development of sexual organs when exposed to a range of MeOH doses by gavage (1.3–5.2 ml/kg or 1023–4090 mg/kg bw according to the Center for the Evaluation of Risks to Human Reproduction (CERHR) calculations (NTP, 2002)) on GD 10 (Youssef et al., 1997). Among mouse strains, CD-1 mice exhibit cephalic neural tube defects (NTDs) with the highest incidence, along with facial and palatal clefts and skeletal defects (Bolon et al., 1994). C57BL/6 mice, on the other hand, primarily exhibit ophthalmic abnormalities, including microphthalmia (smaller eyes) and anophthalmia (absence of the eye), along with facial and palatal clefts and skeletal defects (Rogers et al., 2004). Overall, C57BL/6 mice are more susceptible to the effects of *in utero* MeOH exposure, displaying a much higher incidence of affected pups per litter, more than twice that observed in CD-1 mice. The periods of susceptibility among strains are similar, however, with MeOH administration on GD 7 or 8 resulting in the highest incidence of affected pups.

Comparing the potency of MeOH to the related alcohol, EtOH, in whole embryos of various strains of mouse exposed in culture for 24 hours, for some developmental parameters there is no distinct difference in embryopathies caused by these two alcohols at a similar molar concentration. However, for other parameters, it appears that EtOH reduces anterior neuropore closure, turning and somite development by 50 mM while MeOH reduces these parameters at or above 100 mM, suggesting EtOH is more potent (Figure 7.15). In rat embryos, EtOH appears to be more potent than MeOH, as EtOH reduces crown-rump length, head length, somite development, and protein content by 50–100 mM, while MeOH reduces these parameters at or above 100 mM (Figure 7.16). Some outcomes assessed in the same strain by the same lab exhibited differences, as observed with protein content measured in Sprague-Dawley rat embryos (Figure 7.16). There are no remarkable species differences between mouse and rat embryos exposed to MeOH for 24 hours in culture except for embryolethality,

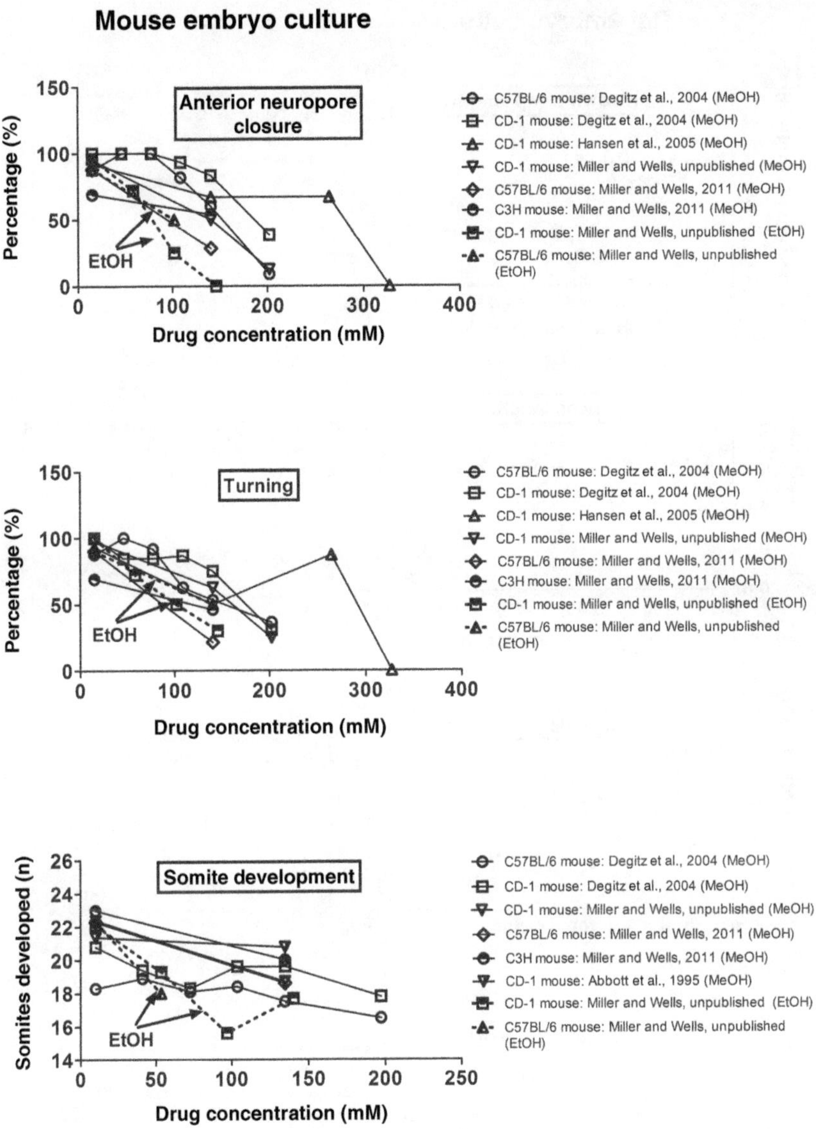

FIGURE 7.15 Comparison of outcomes in mouse embryos exposed to MeOH or ethanol (EtOH) for 24 hours in culture. EtOH studies are represented by dashed lines, and MeOH studies are represented by solid lines. For unpublished studies by Miller and Wells, each group consisted of 11–40 embryos.

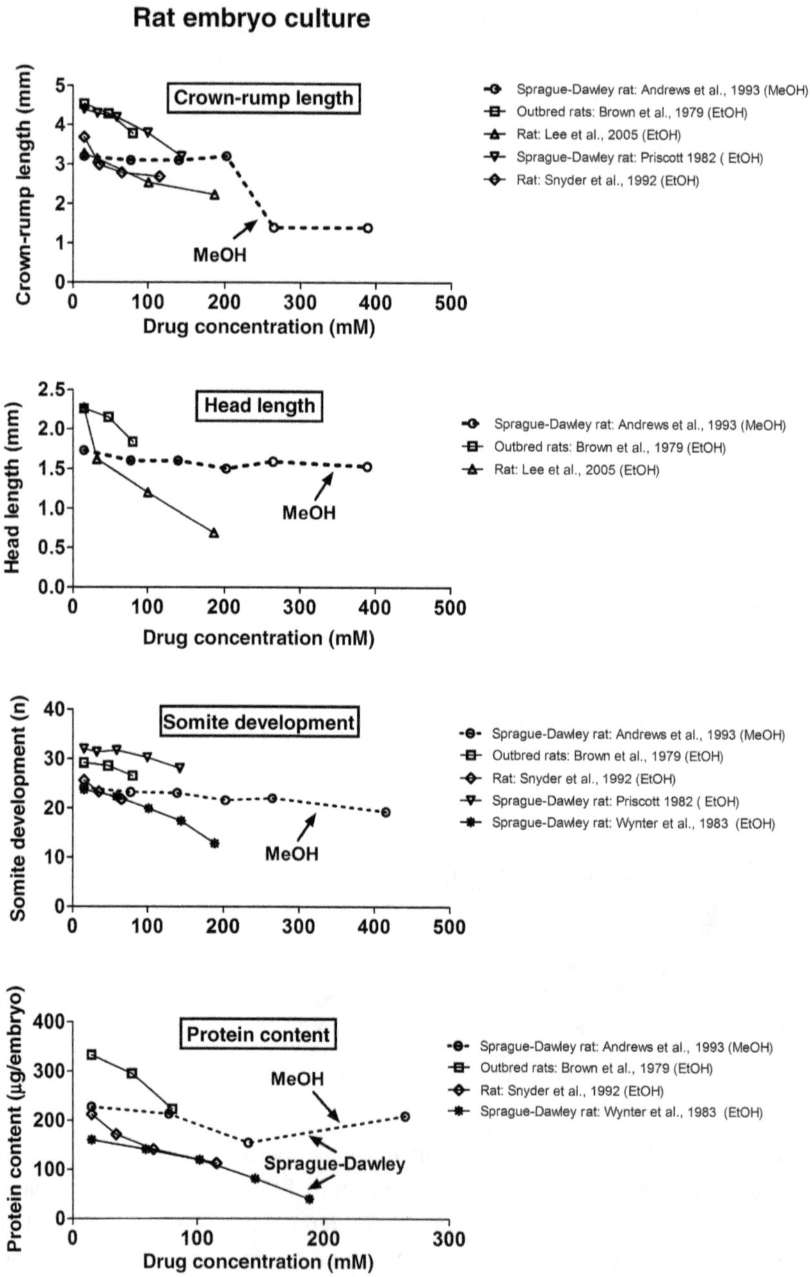

FIGURE 7.16 Comparison of outcomes in rat embryos exposed to ethanol (EtOH) or MeOH for 48 hours in culture. MeOH studies are represented by dashed lines, and EtOH studies are represented by solid lines.

whereby mouse embryos die above 250 mM MeOH, while rat embryos can survive above 350 mM (Figure 7.17).

Comparisons are simpler in a whole embryo culture model where the xenobiotic concentration and duration of exposure can be controlled, whereas comparing the potency of MeOH and EtOH *in vivo* is more difficult. Prenatal exposure of rodents to both alcohols causes birth defects, the incidence of which increases with higher doses, with the anomalies varying depending on the time of exposure as well as the peak blood alcohol concentration (BAC) achieved (Becker et al., 1996) (Table 7.8). Human prenatal EtOH exposure results in a characteristic spectrum of anomalies, collectively termed FASD, which includes decreased fetal weight, distinct craniofacial malformations, and

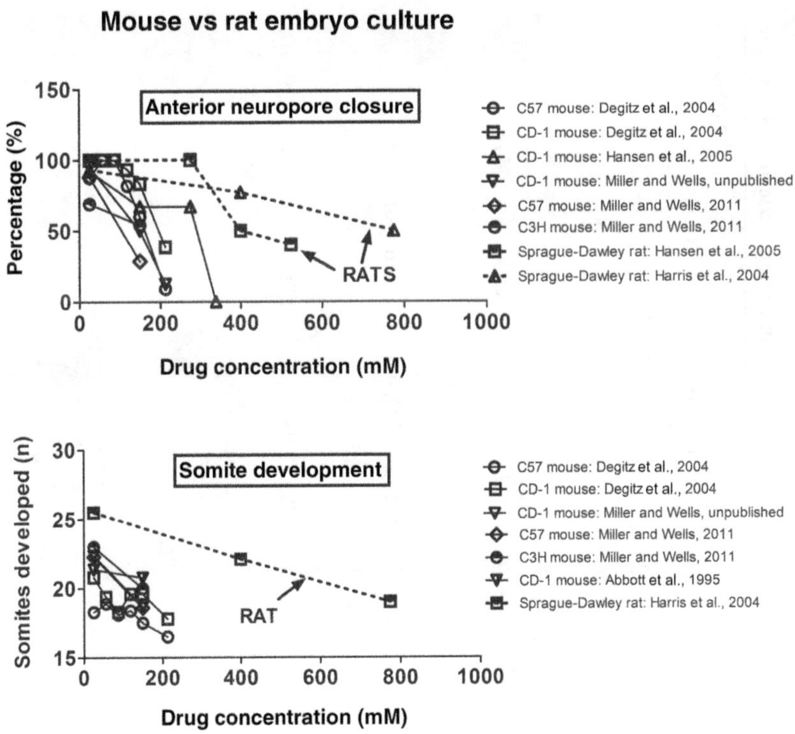

FIGURE 7.17 Comparison of outcomes in mouse and rat embryos exposed to methanol (MeOH) for 24 hours in culture. The rat study is indicated by a dashed line, and the remaining studies employed mice. For unpublished studies by Miller and Wells, each group consisted of 11–40 embryos.

TABLE 7.8 Strain-Specific Malformations Following *in utero* EtOH Exposure in Mice

Model	ROA	Dosing Regimen	BAC	Exposure Period	Analysis Date	Outcome	Citation
C57BL/6	i.p.	2 × 2.9 g/kg (23.7%), 4 h apart	N/D	GD 8	GD 17	↓ crown-rump length, brain, and whole body volume (MRI)	Parnell et al. (2009)
C57BL/6, ICR		4.8 g/kg (25%)	N/D	GD 8	GD 17	↑ resorptions and fetal weight (C57Bl/6 > ICR)	Weston et al. (1994)
C57BL/6		2 × 2.9 g/kg (23.7%), 4 h apart	220 mg/dl	GD 8	GD 8, 10, 14	↑ facial malformations, resorptions	Kotch and Sulik (1992)
C57BL/6, DBA		2 × 3.97 g/kg (25%), 4 h apart	600–640 mg/dl	GD 7, 8, 9, 10, 11, 12, or 13	GD 18	↑ fetal mortality, microphthalmia and digital malformations (C57Bl/6), ↑ ophthalmic defects (DBA), ↓ fetal brain weight	Fukui et al. (1990)

C57BL/6 mouse		2 × 2.9 g/kg (25%), 4 h apart or 5.8 g/kg (25%)	300–500 mg/dl	GD 7, 8, 9, or 10	GD 18	↑ facial malformations (day 7, 8), limb defects (day 9, 10)	Webster et al. (1983)
CBA, C3H, C57BL/6	Liquid diet	20% in water	43–140 mg/dl	Throughout gestation	GD 18	↑ prenatal death, malformations, ↓ fetal weight (CBA > C3H > C57Bl/6)	Chernoff (1980)
C57BL/6		25% in water	N/D	GD 5–10	GD 19	↑ resorption rate, malformations (skeleton, eye, head, heart, abdominal/urogenital)	Randall et al. (1977)

GD, gestational day; BAC, blood alcohol concentration; i.p., intraperitoneal; N/D, not determined; MRI, magnetic resonance imaging.

neurobehavioral deficits (Jones and Smith, 1973). On the other hand, while prenatal MeOH exposure is developmentally toxic in mice and rats, the mechanism is unclear, and it is not known whether MeOH is developmentally toxic in humans. There are several confounding factors in attempting to compare the embryopathic potencies of EtOH and MeOH *in vivo*. In some studies, the maternal peak BAC achieved in the model may not have been measured, let alone concentrations in embryonic or fetal tissues, so it is difficult to compare one alcohol and dose to another. Additionally, mice of different strains exhibit varied susceptibility to the same dose of drug such that their dose–response curves are shifted, with one strain of mouse being more or less susceptible than another strain under the same experimental conditions (Chernoff, 1980; Weston et al., 1994). Furthermore, dosing regimens, routes of administration, as well as outcomes measured differ across studies, precluding direct comparisons. *In vivo* studies comparing the same molar equivalent dose of EtOH and MeOH in the same strains and species under the identical conditions measuring the same developmental endpoints and at least the peak maternal BAC, if not embryonic or fetal tissue concentrations, would be useful in determining more definitively the relative teratological potencies of these two alcohols.

7.3.3 Neurodevelopmental Effects

FASD in humans resulting from EtOH exposure during pregnancy may include neurodevelopmental effects, such as cognitive and other behavioral deficits (Jones, 2011). Behavioral deficits as a result of *in utero* MeOH exposure remain to be definitively determined owing to the lack of reports in humans and investigations in animal models. Only Long-Evans rats given either MeOH in drinking water (GDs 15–17 or 17–19) or inhaled MeOH (GD 7–19) during pregnancy have been studied for postnatal behavioral and cognitive deficits (Infurna and Weiss, 1986; Stanton et al., 1995) (Table 7.9). The first study monitored litter parameters such as litter size, fetal weight, and infant mortality, as well as eye opening, suckling, and nest-seeking (homing) behavior. Pups exposed to MeOH required more time to suckle on postnatal day (PND) 1 and also

TABLE 7.9 Species- and Strain-Specific Neurodevelopmental Effects of *in utero* MeOH Exposure in Rats and Non-human Primates

Model	Strain	ROA	MeOH Dose	Dosing Period	Assessment Parameter(s)	Outcome	Citation
Rat	Long-Evans	MeOH in drinking water, fed *ad libitum*	2% MeOH solutions	GD 15–17 or 17–19 (average 2.5 g/kg/day)	Litter size, birth weight, infant mortality, eye opening, suckling, nest-seeking (homing) behavior (PND 10)	*PND 1*: MeOH-exposed pups required longer than controls to begin suckling *PND 10*: MeOH-exposed pups required more time to locate nesting material from their home cages	Infurna and Weiss (1986) and NTP (2004)
Rat	Long-Evans	Inhalation	15,000 ppm, 7 h/day	GD 7–19	Offspring mortality and body weight (PND 1, 3), motor activity (PND 13–21, 30, 60), olfactory learning (PND 18), behavioral thermoregulation (PND 20–21), T-maze learning (PND 23–24), acoustic startle response (PND 24, 60), reflex modification audiometry (PND 60), pubertal landmarks (PND 31–56), passive avoidance (PND 72), visual-evoked potentials (PND 160)	↓ fetal body weight on PND 1. No other test revealed significant effects of MeOH administration	Stanton et al. (1995) and NTP (2004)

(*continued*)

TABLE 7.9 (*Continued*)

Model	Strain	ROA	MeOH Dose	Dosing Period	Assessment Parameter(s)	Outcome	Citation
Nonhuman primate	*Macaca fascicularis*	Inhalation	200, 600, or 1,800 ppm, 2.5 h/day	7 days/week during a premating and mating period (about 180 days) and throughout the entire gestation period (about 168 days)	First 9 months of life, infant neurobehavioral development (neonatal behavior, early reflex responses, infant gross motor development, spatial memory, and social behavior)	*Visually directed reaching test (sensorimotor development)*: male performance ↓ in all treated groups *Fagan test of infant intelligence (information processing, attention, and visual memory function)*: all treatment groups failed to show preference for novel stimuli, unlike controls No other tests revealed MeOH effects	Burbacher et al. (1999) and NTP (2004)

ROA, route of administration; GD, gestational day (if no other indication, plug = GD 1); PND, postnatal day.

required more time to locate nesting material from their home cages on PND 10, suggesting a delay in normal behavioral development. The second study, where rats were given a chronic high dose (15,000 ppm 7 hours/day, GDs 7–19) of inhaled MeOH, also looked at litter parameters, and in addition employed a variety of behavioral and cognitive tests for motor activity, olfactory learning, behavioral thermoregulation, T-maze learning, acoustic startle response, reflex modification, pubertal landmarks, passive avoidance, and visual evoked potentials. The only observed effect of MeOH was a decrease in fetal body weight, while no effects were observed with any of the behavioral tests employed. A single nonhuman primate study was conducted in the primate model *Macaca fascicularis*, where pregnant monkeys were exposed to a range of concentrations of inhaled MeOH throughout the pre-mating and entire gestational period (Burbacher et al., 1999). Postnatal behavior was followed for 9 months after birth, employing neurodevelopmental behavioral tests for early reflex response, motor development, spatial memory, and social behavior. *In utero* MeOH exposure caused a decrease in sensorimotor development only in male progeny in all treated groups. MeOH exposure also demonstrated a failure for fetuses, in all treatment groups and regardless of sex, to show preference for novel stimuli as tested by the Fagan Test of Infant Intelligence. No other tests were affected by *in utero* MeOH exposure. Results of this study must be interpreted with caution because of the small sample size used to assess the neurobehavioral outcomes, as random fluctuations could potentially account for the statistical differences observed. On the other hand, sensorimotor development is a complex developmental process, and the presence or absence of an observed effect at this early stage does not preclude a delayed onset neurotoxic effect later in life.

7.3.4 Carcinogenic Potential

Although the acute toxicity of MeOH in humans is well described, its carcinogenic potential, similar to its teratogenic potential, is unknown. It is similarly unknown which chemical species, whether MeOH and/or any of its metabolites, might be a proximal toxicant. Formaldehyde itself is a known carcinogen that causes nasopharyngeal cancer in humans and squamous cell carcinomas of the respiratory epithelium of rats and mice (Kerns et al., 1983; IARC, 2006), but this species is rapidly metabolized by ADH3 to FA, and increases in formaldehyde levels were not observed in body fluids or tissues following acute high-dose MeOH exposures in early studies with formaldehyde detection limits of 25 μM (Makar and Tephly, 1977; McMartin et al., 1977; McMartin et al., 1979). Formaldehyde is a metabolic intermediate endogenously produced from amino acid metabolism and exists at concentrations of about 0.1 mM in human blood (IARC, 2006). In addition to rapid metabolism to FA, formaldehyde can be sequestered as macromolecular adducts to endogenous nucleophiles such as DNA, lipid, and protein. Recent formaldehyde inhalation studies using radio-labeled formaldehyde reported that exogenous formaldehyde DNA adducts were observed only in nasal DNA, and that the exogenous adduct levels did not exceed the levels of endogenous formaldehyde DNA adducts, suggesting that additional mechanism(s) in addition to DNA adduct formation are involved in formaldehyde-dependent naso-pharyngeal cancer (Lu et al., 2010; Moeller et al., 2011). Exposure of Sprague-Dawley rats to MeOH (20,000 ppm) or formaldehyde (1500 ppm) in drinking water for 7 days failed to increase the basal levels of the major formaldehyde-DNA adduct N^6-hydroxymethyldeoxyadenosine (N^6-HOMe-dAdo) in leukocyte or hepatocyte DNA (Wang et al., 2008). Another potential indirect mechanism of MeOH-initiated carcinogenesis is via ROS-mediated oxidative damage to DNA. MeOH could promote ROS formation directly via a free radical intermediate, or indirectly via mechanisms including the activation and/or enhancement of ROS-producing NADPH oxidases (NOXs), as has been reported for EtOH (Dong et al., 2010). Several studies have reported free radical production during MeOH biotransformation (Skrzydlewska et al., 2000; Castro et al., 2002; Paula et al., 2003), with the most direct evidence being the direct detection of α-(4-pyridyl-1-oxide)-*N-tert*-butylnitrone (POBN)-hydroxymethyl radical adducts in bile and urine from rats acutely intoxicated with MeOH (Kadiiska and Mason, 2000).

Oxidatively damaged DNA has been implicated for decades as a causal factor in carcinogenesis (Ames, 1989), the most commonly measured lesion being 8-hydroxyguanine, or its physiologically prevalent keto form, 7,8-dihydro-8-oxoguanine, commonly termed 8-oxo-2′-deoxyguanine (8-oxodG). 8-OxodG is the most abundant pro-mutagenic oxidation product of guanine, yielding G-to-T transversion mutations that could activate oncogenes or inactivate tumor suppressor genes linked to the development of cancers (Hsu et al., 2004; Klaunig and Kamendulis, 2004). Genetically altered mice with deficiencies in DNA glycosylases that protect against G-to-T transversions provide strong evidence of a causal role for oxidatively damaged DNA in tumorigenesis (Russo et al., 2004; Xie et al., 2004; Kinoshita et al., 2007). Dimethylarsinic acid strongly increases 8-oxodG levels and carcinogenicity in lungs of *Ogg1* knockout (KO) mice (Kinoshita et al., 2007). In double mutant mice deficient in *Ogg1* and *Myh*, 8-oxodG accumulates in lung and small intestine, and these organs have multi-fold increases in cancer incidence with a high frequency of G-to-T transversion mutations that activate the K-ras oncogene in lung cancers (Xie et al., 2004). Deficiencies in the repair of 8-oxoG have also been suggested to be risk factors for the development of human lung cancer (Paz-Elizur et al., 2003; Mambo et al., 2005).

There are no human data regarding the carcinogenic potential of MeOH, and only four rodent studies, which have reported conflicting results (Table 7.10). Inhalation studies in mice and rats performed by the NEDO (1987) in Japan concluded that carcinogenic effects were not evident following chronic exposure to MeOH, and that there were no differences in the incidence of leukemias or lymphomas. The lack of developmental effects in nonhuman primates is not entirely consistent with the published evidence of developmental toxicity discussed earlier in Section 7.3.2. The NEDO studies were initially difficult to assess as they were available only as summary reports, but more complete English translations are now available from the Methanol Institute (www.methanol.org). An extensive chronic carcinogenic study performed in Sprague-Dawley rats by the Ramazzini Foundation (Soffritti et al., 2002) exposed rats to a range of MeOH concentrations in drinking water and reported dose-dependent increases in lympho-immunoblastic lymphomas and ear duct carcinomas. However, the Ramazzini Foundation studies have been questioned and its results are not widely accepted (NTP, 2004; Cruzan, 2009). First, its methodology does not

TABLE 7.10 Summary of Methanol Carcinogenicity Studies in Animals

Model	Strain	ROA	MeOH Dose	N	Confounders	Outcome	Citation
Mouse	B6C3F1	Inhalation	0, 10, 100, 1,000 ppm, 19.1 h/day for 72 weeks	52 M 53 F	No initial peer review	No MeOH related carcinogenic effects	NEDO (1985)
Mouse	Swiss Webster	Oral	0.222, 0.444, 0.889% in drinking water 6 days/week for lifespan	25 M 25 F	No initial peer review Nno concurrent negative controls Mmalonaldehyde carcinogenicity study with MeOH-only controls Colony not specific-pathogen-free High incidence of pneumonia in study Origin of lymphomas not reported	↑ incidence of total malignant lymphoma in high-dose female and mid-dose male compared to historic untreated controls but also reported incidences are within normal range of occurrences in Eppley Swiss mice No apparent dose response relationship is evident for MeOH exposure and lymphomas when all treatment groups are considered	Apaja (1980)
Mouse	Swiss Webster	Dermal	0.05 ml containing 0.625% H_2O 3 × weekly	40 F	Same as above oral study	No increase in tumor incidence	Apaja (1980)

Rat	Sprague-Dawley	Oral	0, 500, 5,000, 20,000 ppm in drinking water for 104 weeks; observed until death	100 M 100 F	No initial peer review Nonstandard study design Colony not specific-pathogen-free High incidence of lung pathology in all study groups Pathology reporting discrepancies	↑ of all "hemolymphoreticular" tumors combined in high-dose females ↑ in ear duct carcinoma of high-dose males	Soffritti et al. (2002)
Rat	F334	Inhalation	0, 10, 100, 1,000 ppm, 19.5 h/day for 104 weeks	52 M 52 F	No evaluation of dose–response relationships No initial peer review	NEDO reported no increases in tumour incidence U.S. EPA (2009) reanalysis: ↑ in combined lung adenoma and carcinoma in high dose male; 1 lung carcinoma in MeOH group versus 0 in control; ↑ in pheochromocytomas in high-dose females compared to NTP historical controls but not in concurrent controls	NEDO (1985)

EPA, Environmental Protection Agency; F, female; M, male; N, number of animals; NEDO, New Energy Development Organization (Japan); NTP, U.S. National Toxicological Program.

allow for an accurate estimate of the actual dose of MeOH consumed by the rats in drinking water. Total water consumption was measured only weekly, rather than daily, per cage, where multiple rats were housed, which may inappropriately average individual variability. The classification of the results made their interpretation difficult, as the diagnostic terms used did not match standard pathological guidelines. Of the classifications made by the pathologists at the Ramazzini Foundation, nearly half did not match those reviewed by the National Toxicology Program (NTP, 2004). There were also discrepancies with the study's control data in comparison to historical control data, where the incidences of total cancers derived from blood-forming cells were consistently approximately four times higher. Furthermore, their report neglected to acknowledge the high incidence of early mortality (>80% of rats in study) and lung pathologies present in nearly all dose groups, which likely contributed to the formation of the reported lympho-immunoblastic lymphomas. Additionally, the report lacked pertinent information that could help the interpretation of results, such as the aforementioned limited data on water consumption and limited information concerning lung pathology. These limitations diminished the validity of the study, and its results would require replication to corroborate the conclusions made in the report. One additional report in the literature is a graduate thesis examining the carcinogenic potential of malonaldehyde in Swiss Webster Mice, in which MeOH was utilized as a vehicle control, although unfortunately there were no concurrent untreated controls in this study (Apaja, 1980). Mice were exposed to MeOH in drinking water at doses of approximately 550, 1000, or 2000 mg/kg/day six times per week until their spontaneous death. The incidence of malignant lymphomas in MeOH-treated mice was higher than the overall incidence in historical controls for high-dose females and mid-dose males, but the author concluded that these incidences were "within the normal range of occurrence of malignant lymphomas in Eppley Swiss mice." The mouse colony was not maintained under specific pathogen-free (SPF) conditions, and there was a high incidence of pneumonia reported (8–28%), which may be a confounding variable.

7.4 OXIDATIVE STRESS

7.4.1 Oxidative Stress Mechanisms

7.4.1.1 Embryonic Drug Exposure and Reactive Oxygen Species (ROS) Formation Maternal elimination of a teratogen can be an important regulator of the amount reaching the embryo, so maternal pathways such as CYP-catalyzed hydroxylation reactions and UDP-glucuronosyltransferase-catalyzed drug conjugation are important determinants of embryonic teratogen exposure (Wells et al., 2005). However, because ROS, and particularly hydroxyl radicals, are highly reactive and unlikely to escape the cell in which they are formed, let alone the tissue or organ, maternal pathways of ROS formation are unlikely to contribute to embryonic ROS levels, which are determined by proximate, embryonic pathways. The risk of embryopathies will likely be determined by a balance among (1) the maternal pathways of teratogen elimination and (2) embryonic pathways of ROS formation and detoxification, and repair of ROS-mediated oxidative macromolecular damage to cellular macromolecules such as DNA, protein, and lipids. When an imbalance in these pathways occurs, teratogenesis can result even at therapeutic drug doses or maternal plasma concentrations, or at exposures to levels of environmental chemicals generally considered to be safe.

Drugs and environmental chemicals can enhance ROS formation via a number of mechanisms that are not necessarily mutually exclusive for a given xenobiotic. These mechanisms include (1) enzymatic bioactivation to a free radical intermediate, catalyzed by cytochromes P450s (CYPs), prostaglandin H synthases (PHSs), and lipoxygenases (LPOs), among others (Wells et al., 2009b, 2010) (Figure 7.18); (2) superoxide formation during the metabolism of substrates such as EtOH by CYP2E1 (Koop, 2006); (3) redox cycling of catechol metabolites (Wang et al., 2010); (4) interference with the mitochondrial electron transport chain, producing superoxide (Maritim et al., 2003); and (5) activation and/or induction of enzymes such as the NOXs that form superoxide and/or hydrogen peroxide (Lambeth, 2004; Brown and Griendling, 2009; Jiang et al., 2011) (Figure 7.19).

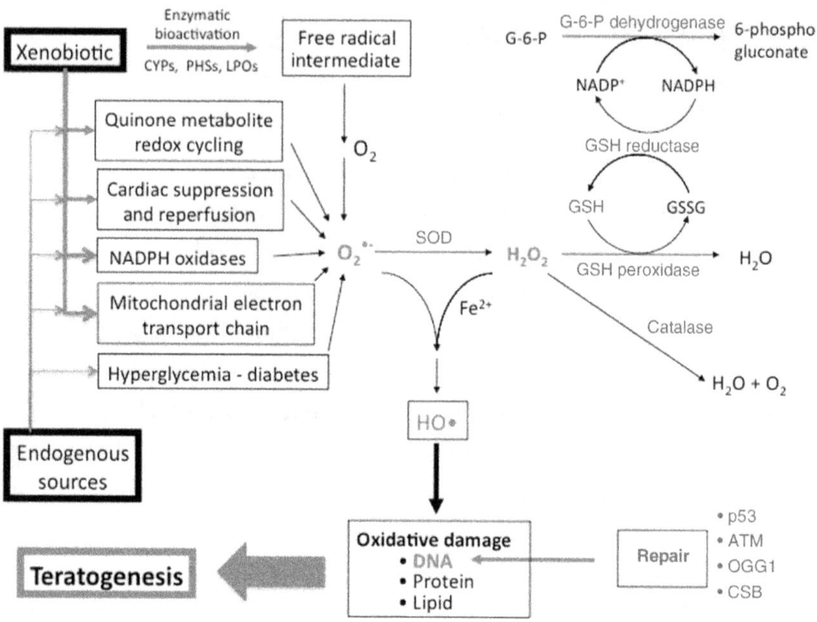

FIGURE 7.18 Biochemical pathways for endogenous and xenobiotic-enhanced formation and detoxification of reactive oxygen species (ROS), and repair of oxidatively damaged cellular macromolecules. ROS include superoxide ($O_2•^-$), hydrogen peroxide (H_2O_2), and hydroxyl radicals (HO•). Teratogenesis is postulated to result from embryonic macromolecular damage and/or ROS-mediated alterations in embryonic signal transduction. If embryonic ROS formation exceeds the proximal capacity for ROS detoxification and/or repair of cellular macromolecules, this imbalance can result in enhanced teratogenesis, even at a therapeutic drug concentration or generally "safe" exposure level for an environmental chemical. Abbreviations: ATM, ataxia telangiectasia mutated protein; CSB, Cockayne syndrome B protein; CYPs, cytochromes P450; Fe, iron; G-6-P, glucose-6-phosphate; GSH, glutathione; GSSG, glutathione disulfide; LPOs, lipoxygenases; $NADP^+$, nicotinamide adenine dinucleotide phosphate; OGG1, oxoguanine glycosylase 1; PHSs, prostaglandin H synthases, SOD, superoxide dismutase. *Source*: Modified from Wells et al. (2009b).

7.4.1.2 Signal Transduction ROS are widely implicated in highly regulated cellular signal transduction pathways, which are selective to different cell types and their subcellular organelles (Wells et al., 2009b) (Figure 7.20). ROS signaling has been linked to numerous pathways involved in cellular proliferation, differentiation, migration, and apoptosis (Thannickal and Fanburg, 2000). ROS-mediated signal transduction has been attributed to hydrogen peroxide, which is less reactive and

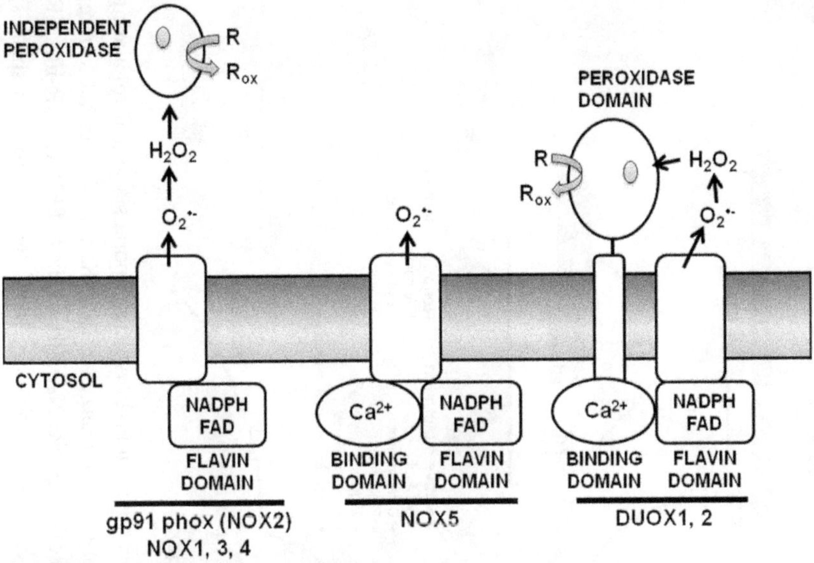

FIGURE 7.19 NADPH oxidase-mediated ROS formation. Abbreviations: NADPH, nicotinamide adenine dinucleotide phosphate; NOX, NADPH oxidase; FAD, flavin adenine dinucleotide; MPO, myeloperoxidase; DUOX, dual oxidase; R, extracellular substrates. *Source*: Modified from Lambeth (2004).

has greater diffusibility than superoxide or hydroxyl radicals, allowing it to selectively oxidize sulfhydryl groups of specific cysteine residues on proteins resulting in several reversible modifications including the formation of protein–protein (Pr–Pr) and GS–protein disulfides (GS–Pr, mixed disulfides) (Thannickal and Fanburg, 2000; Janssen-Heininger et al., 2008). At physiologically relevant concentrations of hydrogen peroxide, these modifications are reversible and constitute a control mechanism of protein function, whereas higher exposures could lead to excessive and irreversible S-oxidation resulting in loss of protein function and pathological consequences. Major cellular sources of ROS include the mitochondrial electron transport chain, LPOs, as well as NOXs (Lander, 1997). NOX enzymes are commonly known for their role in phagocytes for initiating the "respiratory burst" by producing high concentrations of superoxide and hydrogen peroxide. Homologs were later found to be expressed ubiquitously in nonphagocytic cells, although they produce more physiological levels of ROS involved in signal transduction (Lambeth, 2004; Brown and Griendling,

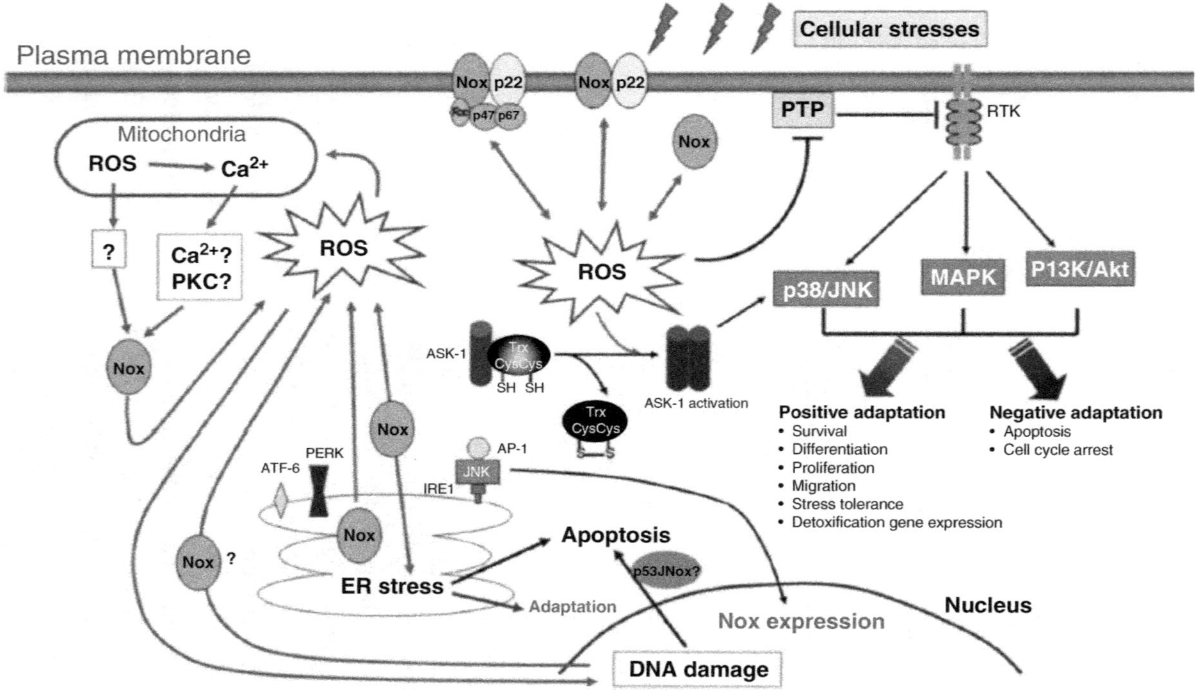

FIGURE 7.20 ROS-mediated signal transduction. Abbreviations: AP-1, activator protein-1; ASK-1, apoptosis signal-regulating kinase-1; ATF-6, activating transcription factor 6; Ca^{2+}, calcium; IRE1, inositol-requiring enzyme 1; JNK, jun N-terminal kinase; MAPK, mitogen-activated protein kinase; NADPH, nicotinamide adenine dinucleotide; NOX, NADPH oxidase; PERK, PKR-like ER kinase; PI3K, phosphoinositide 3-kinase; PKC, protein kinase C; PTP, protein tyrosine phosphatase; ROS, reactive oxygen species; RTK, receptor tyrosine kinase; Trx, thioredoxin; Cys, cysteine; ?, unknown factors. *Source*: Reproduced from Jiang *et al*., (2011).

2009). ROS can modulate signaling pathways involving protein kinases and transcription factors such as nuclear factor transcription factor kappa B (NF-κB), p53, mitogen-activated protein kinases (MAPK), and protein kinase C (Halliwell and Gutteridge, 2007). ROS-initiated transcription factor activation can alter cellular gene expression leading to the downstream cellular response (Brown and Griendling, 2009). A specific example of xenobiotic-initiated ROS-mediated signal transduction altering embryonic development is *phenytoin*, an antiepileptic drug given during pregnancy to mitigate seizure occurrence. Our lab has shown the involvement of the NF-κB signaling pathway in phenytoin-initiated embryopathies in culture (Kennedy et al., 2004). Using antisense oligonucleotides, inhibition of the downstream NF-κB signaling cascade blocked embryopathies, suggesting the involvement of ROS-initiated NF-κB signaling in phenytoin-initiated embryopathies. The NF-κB family of transcription factors regulates the expression of many genes involved in development, as well as immunity and the inflammatory response (Baeuerle and Baltimore, 1996). Another example is *thalidomide*, a drug used for its sedative-hypnotic effects in pregnant women, and is still in clinical use for the treatment of leprosy and multiple myeloma because of its strong immunomodulatory, anti-inflammatory, and antiangiogenic properties. When taken during the third to eighth week of pregnancy, thalidomide can initiate birth defects predominantly involving the limbs (phocomelia is the most well known) but can also affect the ears, eyes, heart, kidneys, and other internal organs (Knobloch and Ruther, 2008). The mechanisms by which thalidomide initiates birth defects are not clear; however, several mechanisms have been suggested, including oxidative stress and alterations in signal transduction pathways. Thalidomide has been demonstrated to suppress numerous survival signaling pathways including the canonical Wnt/β-catenin pathway (Knobloch et al., 2007) and Akt signaling, while upregulating phosphatase and tensin homolog (PTEN) and stimulating caspase-dependent cell death (Knobloch et al., 2008).

7.4.1.3 Macromolecular Damage Elevated ROS concentrations increase the likelihood that they will react with molecular oxygen to form superoxide via a one-electron reduction, hydrogen peroxide via a second electron reduction or enzymatically catalyzed by superoxide dismutase, or the highly reactive hydroxyl radical via the Fenton reaction with hydrogen peroxide. Hydroxyl radical can oxidatively damage cellular macromolecules including proteins, DNA, and lipids (Halliwell and Gutteridge, 2007).

Protein Oxidation Protein oxidation can impair the function of signal transduction proteins, receptors, and enzymes and subsequently cause secondary damage to other cellular macromolecules. If damage occurs to DNA repair enzymes, levels of DNA damage can increase, while damage to DNA polymerases may decrease replication fidelity (Halliwell and Gutteridge, 2007). Oxidized protein may be recognized as a foreign antigen by the innate immune system and trigger antibody production (Peng et al., 1997). Oxidation of a protein is initiated by the hydroxyl radical-dependent abstraction of the α-hydrogen atom of an amino acid residue to form a carbon-centered radical, which can rapidly react with oxygen to form subsequent radical intermediates that can react with other amino acid residues in the same or a different protein, forming a new carbon-centered radical (Berlett and Stadtman, 1997). All amino acid side chains are susceptible to attack by hydroxyl radicals. A well-studied measure of protein oxidation is protein carbonyl formation. Carbonyl groups can be introduced into proteins either by direct metal-catalyzed oxidation of lysine, arginine, proline, and threonine residues, or by reaction with aldehydes produced during lipid peroxidation, such as 4-hydroxy-2-nonenal (HNE), and malondialdehyde (MDA), or with reactive carbonyl derivatives (Berlett and Stadtman, 1997; Nystrom, 2005). Histidine attacked by the reactive lipid peroxidation aldehyde product HNE can generate an *HNE-histidine adduct* that can be measured as a marker of lipid peroxidation (Uchida and Stadtman, 1992) (Figure 7.21). Oxidized protein products are removed from the cell by increased recognition and degradation by cellular proteases, and loss of developmentally important proteins in the embryo can lead to subsequent embryopathies, which has been observed with phenytoin-initiated protein oxidation (Winn and Wells, 1999; Nystrom, 2005). The presence of elevated levels of carbonylated proteins has been used as a marker of ROS-mediated protein oxidation, and several methods of detection have been developed (Levine et al., 1994).

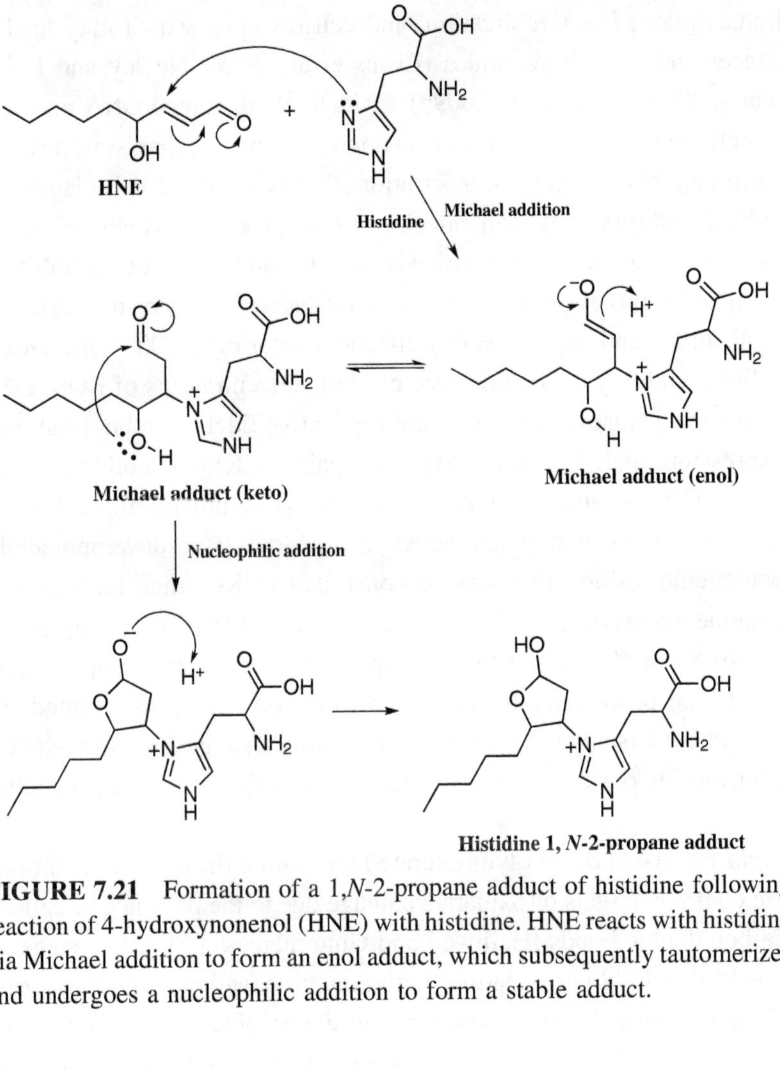

FIGURE 7.21 Formation of a 1,*N*-2-propane adduct of histidine following reaction of 4-hydroxynonenol (HNE) with histidine. HNE reacts with histidine via Michael addition to form an enol adduct, which subsequently tautomerizes and undergoes a nucleophilic addition to form a stable adduct.

DNA Oxidation Exposure of DNA to highly reactive hydroxyl radicals can yield several products, depending where the oxidation occurs, which can accumulate and may contribute to teratological outcomes.

Approximately 20 forms of oxidatively damaged DNA have been identified, the most commonly measured being 8-oxodG (Halliwell and Gutteridge, 2007; Wells et al., 2009b). Consequences of 8-oxodG accumulation in dividing cells may lead to transversion mutations, which can affect the expression and activity of proteins required for normal development and function, and if not repaired, may affect gene transcription, DNA replication, and cell division, which may lead to cancer and/or embryopathies (Evans et al., 2004; Neeley and Essigmann, 2006; Wells et al., 2009b). Oxidatively damaged DNA may also directly initiate embryopathies via nonmutagenic mechanisms, possibly including altered gene transcription (Wells et al., 2010). Hydroxyl radicals alternatively can attack nuclear proteins, which results in the formation of protein radicals which can then bind to DNA to form DNA–protein cross links that can interfere with gene transcription, replication, and repair (Halliwell and Gutteridge, 2007). To ensure cellular viability in the presence of high concentrations of ROS, DNA repair mechanisms exist to ensure replicative fidelity and normal gene expression, including base excision repair (BER), nucleotide excision repair (NER), and mismatch repair (Christmann et al., 2003). In addition to its mutagenic activity, 8-oxodG is a developmentally pathogenic lesion. This was demonstrated in KO mice lacking oxoguanine glycosylase 1 (OGG1), a component of the BER pathway that repairs 8-oxodG. *Ogg1* KO mice exposed *in utero* to methamphetamine and tested postnatally for motor coordination deficits performed significantly worse than wild-type (WT) controls, suggesting 8-oxodG can contribute to postnatal neurodevelopmental deficits (Wong et al., 2008).

Lipid Peroxidation Polyunsaturated fats within the cellular membrane are common targets of oxidative damage due to the presence of carbon-carbon double bonds (Halliwell and Gutteridge, 2007). Three steps are involved in lipid peroxidation: initiation, propagation, and termination. (1) Initiation: lipid peroxidation begins either by addition of a hydroxyl group across a double bond, forcing the electrons to move onto the adjacent carbon forming a carbanion, or more commonly, by hydrogen abstraction creating a lipid radical (Gutteridge, 1984; Halliwell and Gutteridge, 2007). Transition metals such as iron and copper can participate in electron exchange with oxygen to form the hydroxyl radical. (2) Propagation: lipid radicals can stabilize by rearrangement to a conjugated diene or can react with molecular oxygen to give rise to a peroxyl radical (ROO•). This peroxyl radical can abstract a hydrogen atom from an adjacent fatty acid side chain, forming new carbon-centered radicals that can react with oxygen to form new peroxyl radicals. The peroxyl radical can combine with the abstracted hydrogen atom to form a lipid hydroperoxide (ROOH). Cyclic peroxides can form when a peroxyl radical attacks a double bond within the same fatty acid residue. (3) Termination: the chain reaction terminates when two lipid peroxyl radicals combine to produce a nonradical species, or when a radical is halted by binding to antioxidants such as α-tocopherol (Halliwell and Gutteridge, 2007).

Lipid peroxidation can produce DNA-damaging aldehydes such as malondialdehyde (MDA), 4-hydroxynonenal (4-HNE), and F-isoprostanes. The decomposition of lipid peroxides by heating or reaction with metal ions creates a wide variety of cytotoxic products, which can produce more radicals that can initiate further lipid peroxidation (Gutteridge and Quinlan, 1983).

MDA is produced either from the peroxidation of polyunsaturated fatty acids (PUFA s) with more than two double bonds, or enzymatically during the metabolism of eicosanoids. At physiological pH, most MDA exists as the enolate ion, which has low reactivity toward amino groups in proteins (Halliwell and Gutteridge, 2007). At a lower pH, MDA exists as the undissociated enol form in equilibrium with its keto form and exhibits a higher reactivity toward proteins, which can attack residues resulting in intra- and intermolecular cross-links (Esterbauer et al., 1991). MDA can react with DNA, more specifically guanine bases, to create G to T transversions, A to G transitions, C to T transitions, frameshifts, and deletions, with potentially mutagenic consequences (Marnett, 2000). MDA is metabolized to malonic semialdehyde by aldehyde dehydrogenase, and this product is decarboxylated to acetaldehyde, and is finally metabolized to acetic acid again by aldehyde dehydrogenase (Halliwell and Gutteridge, 2007).

The thiobarbituric acid (TBA) assay is a colorimetric assay commonly used to detect MDA, whereby a sample is heated with TBA in acid and a pink color develops, reflecting the formation of thiobarbituric acid-reactive substances (TBARS). Although technically easy to perform, the specificity of this assay has been questioned (Gutteridge and Quinlan, 1983). It has been suggested that much of the MDA measured may actually be generated during the assay workup since the amount of free MDA in most lipid peroxidizing systems is too low to be detected by the assay. As much as 98% of the MDA detected can be generated during the acid heating step of the assay by the decomposition of lipids, which can result in a wide variety of radical-generating toxic products (Gutteridge and Quinlan, 1983). This can be circumvented by measuring MDA directly via high-performance liquid chromatography (HPLC) or gas chromatography–mass spectrometry (GC–MS). Another source of uncertainty in the TBARs assay is that several compounds can react with TBA to form chromogens that absorb at 532 nm. Lastly, assaying human body fluids can detect artifactual MDA produced enzymatically during eicosanoid synthesis (Shimizu et al., 1981).

HNE is formed during the oxidation of $n - 6$ PUFAs, which are fatty acids that contain a double bond at the carbon-6 position, such as linoleic acid and arachidonic acid (Esterbauer et al., 1991; Spiteller, 1998). Basal cellular levels of HNE in healthy tissues are approximately 1 μM or lower; however, under conditions of oxidative stress, HNE concentrations can rise to between 2 and 20 μM which is cytotoxic, leading to inhibition of DNA and protein synthesis, cellular proliferation, and NER (Parola et al., 1999; Feng et al., 2004). HNE can react rapidly with thiol and amino groups on proteins (i.e., histidine, lysine) and amino groups on DNA bases, with guanine being the preferred target (Halliwell and Gutteridge, 2007). HNE reacts with DNA to form an etheno adduct by adding an NH_2 group to the double bond of the aldehyde to yield 1,N^2-propano-2^1-deoxyguanosine adducts (Schaur, 2003; Choudhury et al., 2004) (Figure 7.22). As mentioned previously, HNE and MDA, the end products of lipid peroxidation, are DNA-damaging aldehydes. As such, they may facilitate cancer development in several ways such as (1) mutagenic adduct formation with DNA bases, (2) formation of subsequent ROS during the peroxidation process which may directly oxidize DNA leading to mutagenesis, or (3) oxidation of DNA repair proteins resulting in decreased replication fidelity with a subsequent increase in the incidence of mutations (Halliwell and Gutteridge, 2007).

Isoprostanes (IPs) are formed from PUFAs with at least three double bonds, which include linolenic acid, arachidonic acid (F_2-isoprostanes), eicosapentanoic acid (EPA) (F_3 isoprostanes), and docosahexanoic acid (DHA) (F_4 isoprostanes) (Fam and Morrow, 2003; Roberts and Fessel, 2004). IPs can form isoketals that quickly react with amino groups to form adducts with lysine residues on membrane proteins, resulting in

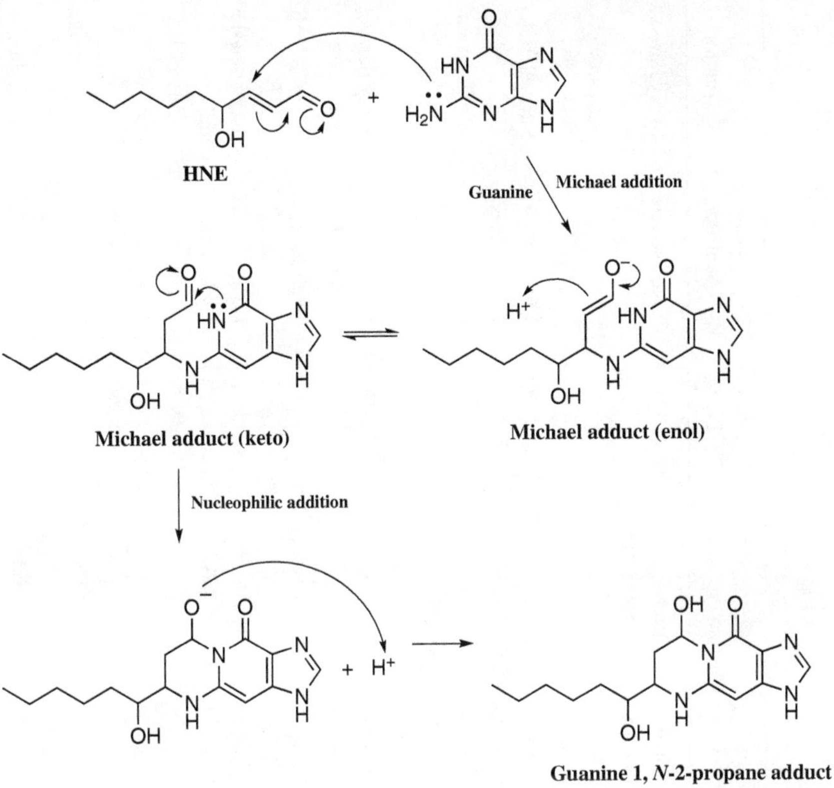

FIGURE 7.22 Formation of a 1,*N*-2-propane adduct of guanine following reaction of 4-hydroxynonenol (HNE) with guanine. HNE reacts with guanine via Michael addition to form an enol adduct, which subsequently tautomerizes and undergoes a nucleophilic addition to form a stable adduct.

protein cross-linking, ultimately damaging the cellular membrane (Poliakov et al., 2004).

7.4.2 Oxidative Stress from Methanol Exposure

7.4.2.1 Evidence for MeOH-Initiated ROS Formation Studies have demonstrated enhanced lipid peroxidation in brain, liver, bile, erythrocytes, urine as well as lymphoid organs of rats intoxicated both acutely and chronically with MeOH (Skrzydlewska and Farbiszewski, 1997b; Skrzydlewska et al., 1998, 2000; Dobrzynska et al., 2000; Parthasarathy et al., 2006a,b) (Table 7.11). Following an acute exposure of rats to MeOH, one group reported an increase in enzymatic and nonenzymatic

TABLE 7.11 Evidence for MeOH-Initiated Oxidative Stress

Model	Strain	ROA	Dose, g/kg	Duration	Probe	Outcome	Effect of Probe	Tissues Analyzed	Citation
Rat	Wistar	i.g.	1.5, 3	Acute	N/A	↑ Lipid peroxidation (TBARS), ↓ GSH-Px, SOD, GSSG-R activities, ↓ [ascorbate]	N/A	Liver, serum	Skrzydlewska and Farbiszewski (1997c)
Rat	Wistar	i.g.	6	Acute	N/A	↑ Lipid peroxidation (TBARS), ↓ GSH-Px, SOD, GSSG-R activities, ↓ sulfhydryl compounds, [ascorbate]	N/A	Liver, erythrocytes, serum	Skrzydlewska and Farbiszewski (1997a)
Rat	Wistar	i.g.	3, 6	Acute	N/A	↓ [GSH], GSH-Px and GSSG-R activities	N/A	Liver, erythrocytes, serum	Skrzydlewska and Farbiszewski, (1997c)
Rat	Wistar	i.g.	3	Acute	U-83836E	↑ Lipid peroxidation (MDA); ↓ GSH-Px, GSSG-R, TAS	↓ lipid peroxidation; ↑ GSH, antioxidative enzymes	Liver, erythrocytes, serum	Skrzydlewska and Farbiszewski (1997d)
Rat	Wistar	i.g.	1.5, 3, 6	Acute	N/A	↑ Lipid peroxidation (TBARS) CuZnSOD, CAT; ↓ GSH-Px, GSSG-R, [GSH], [Ascorbate], TAS	N/A	Liver	Skrzydlewska and Farbiszewski (1997d)

Rat	Wistar	i.g.	6	7 days	N/A	↑ Lipid peroxidation (TBARS), ↓ [GSH]/[ascorbate] *Brain*: ↓SOD, GSH-Px, GSSG-R activity *Liver*: ↑SOD, CAT activity, ↓GSH-Px, GSSG-R activity	N/A	Brain, liver	Skrzydlewska et al. (1998)
Rat	Wistar	i.g.	6	7 days	N/A	Direct detection of MeOH radical, lipid peroxidation, protein carbonyl formation	N/A	Liver	Skrzydlewska et al. (2000)
Rat	Wistar	i.g.	3	Acute	U-83836E, N-Acetylcysteine	↑ Lipid peroxidation (TBARS), ↑ membrane surface charge density	↓ TBARS, membrane surface charge density	Erythrocytes	Dobrzynska et al. (1999)
Rat	Sprague-Dawley	i.p./i.g.	4.5, 7	Acute	POBN (α-(4-pyridyl 1-oxide)-*N-tert*-butylnitrone	Direct detection of a methanol-derived POBN adduct		Bile and urine	Kadiiska and Mason (2000)
Rat	Wistar albino	i.p.	2.37	1, 15, 30 days	N/A	↑ Lipid peroxidation (TBARS), enzymatic/nonenzymatic antioxidants ↑ at 1days; ↓ at 15, 30 days	N/A	Lymphoid organs	Parthasarathy et al. (2006a)

(*continued*)

TABLE 7.11 ***(Continued)***

Model	Strain	ROA	Dose, g/kg	Duration	Probe	Outcome	Effect of Probe	Tissues Analyzed	Citation
Rat	Wistar albino	i.p.	2.37	1, 15, 30 days	N/A	↑ Lipid peroxidation (TBARS), enzymatic/non enzymatic antioxidants ↑ at 1 day; ↓ at 15, 30 days; DNA fragmentation ↑ at 30 days	N/A	Hypothalamus, adrenal gland	Parthasarathy et al. (2006b)

TBARS, thiobarbituric reactive acid substances; SOD, superoxide dismutase; GSH-Px, glutathione peroxidase; GSSG-R, glutathione reductase; MDA, malondialdehyde; TAS, total antioxidant score; CAT, catalase; U-83836E, antioxidant trolox derivative; N/A, not applicable, i.p., intraperitoneal; i.g., intragastric.

antioxidants such as catalase, superoxide dismutase, GSH, and vitamin C, which may be a cellular response to counteract the enhanced ROS formation (Parthasarathy et al., 2006a, 2006b). In contrast, following a longer duration of MeOH exposure, there was a decrease in enzymatic and nonenzymatic antioxidants, which could reflect cellular membrane and macromolecular damage due to excessive oxidative stress.

7.4.2.2 Mechanism of MeOH-Initiated ROS Formation The mechanisms by which MeOH and/or its metabolites enhance ROS formation are yet to be determined. Numerous studies have implicated free radical-initiated, ROS-mediated involvement in the mechanism of toxicity including (1) direct detection of a MeOH radical by electron spin resonance spectrometry, and oxidative protein damage in MeOH intoxicated rats (Skrzydlewska et al., 2000); (2) MeOH-derived adducts to the free radical spin trapping agent, alpha-phenyl-N-tert-butylnitrone (PBN), detected in bile and urine of PBN-pretreated, MeOH-exposed rats (Kadiiska and Mason, 2000); and (3) MeOH embryopathies in rat whole embryo culture are enhanced by the depletion of GSH (Harris et al., 2004). Additionally, MeOH-initiated oxidative stress, as evidenced by the production of the lipid peroxidation product MDA, along with increases in antioxidative enzyme activities, was observed in the lymphoid organs of adult rats (Parthasarathy et al., 2006a). Although the enzyme(s) catalyzing this reaction remain to be determined, PHSs do not appear to contribute, as MeOH embryopathies are not blocked by pretreatment with the PHS inhibitor acetylsalicylic acid (ASA) (Miller and Wells, unpublished). CYP2E1 expression is negligible during the embryonic period, and low compared to adult activity during the fetal period, particularly in rodents (Vieira et al., 1996; Juchau et al., 1998; Hines, 2008), so CYP2E1-mediated superoxide formation would appear to be an unlikely mechanism for embryonic ROS formation, at least in rodents. Preliminary studies suggest that MeOH and/or its metabolites may activate and/or induce the expression of embryonic NOXs that produce embryopathic ROS, as MeOH embryopathies are reduced by pretreatment with the NOX inhibitor diphenyleneiodonium (DPI) (Miller and Wells, unpublished). This would be consistent with the NOX-dependent ROS mechanism previously reported for the related alcohol EtOH (Dong et al., 2010). Additional potential mechanisms, such as interference with the mitochondrial electron transport chain, are yet to be investigated.

7.4.3 Teratogenicity of Methanol and Comparisons to Ethanol

7.4.3.1 Genetic Modulation of Catalase Genetically modified mice with altered catalase activity were used to evaluate the role of catalase in embryo culture, which removes potentially confounding maternal factors, and *in vivo*. Mutant acatalasemic mice (aCat) with reduced catalase activity were compared to their C3H WT controls, and transgenic mice expressing human catalase (hCat) with enhanced catalase activity were compared to their C57BL/6J WT controls (Miller and Wells, 2011).

In embryo culture, MeOH was embryopathic in both the WT C57BL/6J and C3H strains although the range and severity of embryopathies were greater in the C57BL/6J strain (Miller and Wells, 2011). hCat embryos with enhanced catalase activity were protected from MeOH-initiated decreases in anterior neuropore closure, turning, and somite development. In contrast, MeOH-exposed catalase-deficient aCat embryos exhibited reduced anterior neuropore closure and head length, which were not observed in MeOH-exposed WT controls (Figures 7.23 and 7.24). The respectively contrasting protection and increased embryopathies in hCat and aCat embryos compared to their WT controls suggest that ROS contribute to the mechanism of MeOH teratogenicity, and that embryonic catalase, despite constituting only about 2–5% of maternal activity, plays a developmentally protective role. This interpretation is consistent with similar embryo culture and *in vivo* studies of the ROS-initiating teratogen phenytoin (Abramov and Wells, 2011). The results also suggest that the alternative peroxidative activity of catalase in converting MeOH to formaldehyde, at least in embryonic tissues, does not play a measurable role in modulating embryopathic risk.

In the same genetically modified mice, the effects of altered catalase activity in the more complicated *in vivo* system were more difficult to interpret. The C57BL/6J WT controls for the hCat dams treated with MeOH (total dose of 4 g/kg i.p. on GD 8) had offspring

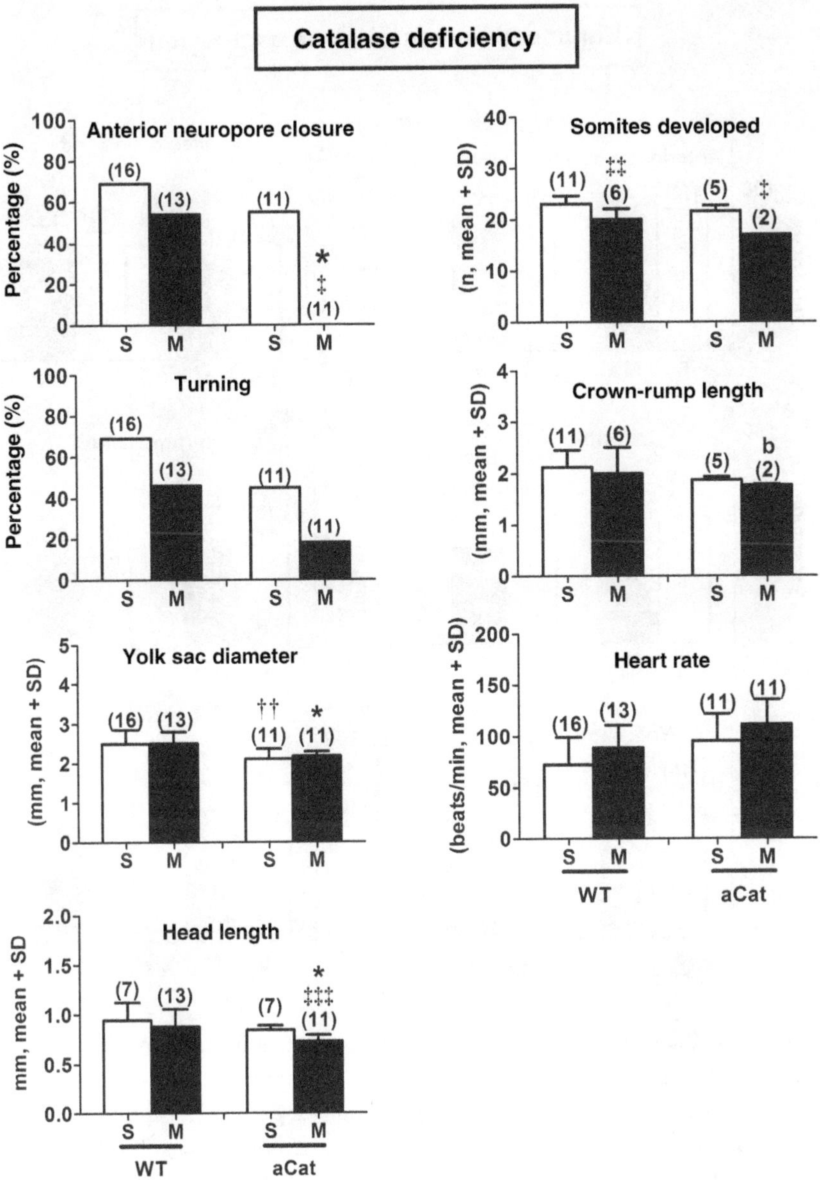

FIGURE 7.23 Effect of MeOH on embryopathies in mice with a catalase deficiency in embryo culture. Double daggers indicate difference from saline treated embryos of same genotype (‡ $p < 0.05$, ‡‡ $p < 0.01$, ‡‡‡ $p < 0.0001$). Asterisk indicates difference between genotypes for the same treatment group (* $p < 0.05$). Letter b indicates $p = 0.05$ using a *t*-test.

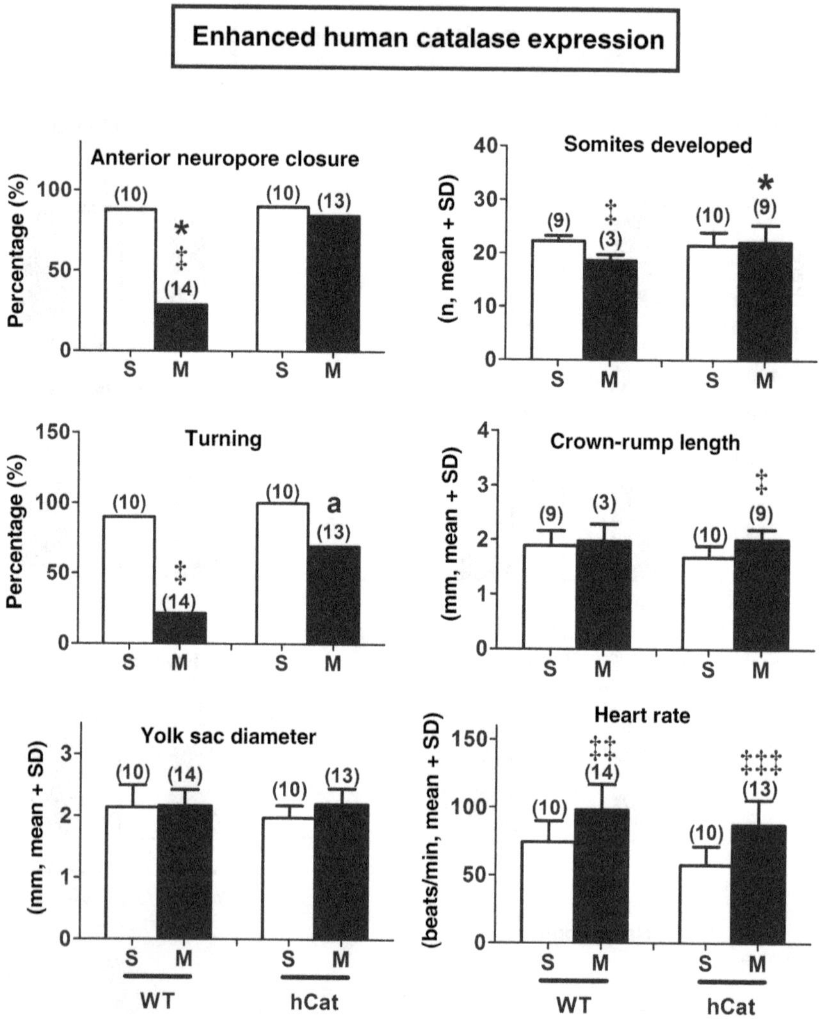

FIGURE 7.24 Effect of MeOH on embryopathies in mice overexpressing human catalase. Double daggers indicate difference from saline treated embryos of the same genotype (‡ $p<0.05$, ‡‡ $p<0.01$, ‡‡‡ $p<0.0001$). Asterisk and letter indicates difference between genotypes for the same treatment group (* $p<0.05$; a, $p=0.05$). *Source*: From Miller and Wells (2011).

with an increased incidence of ophthalmic abnormalities and cleft palates; however, these incidences of anomalies were not significantly different from those seen in hCat mice although there was a trend for protection (Siu et al., 2013) (Figure 7.25). On the other hand, both the aCat mice and their WT C3H controls were relatively resistant to MeOH teratogenicity, the first strain of mouse to be reported as such, precluding an evaluation of the effect of catalase deficiency upon MeOH teratogenesis. The resistance of the aCat and C3H WT controls could not be explained by pharmacokinetic determinants since the pharmacokinetic profiles for MeOH and FA in the aCat, hCat, and WT mice were similar, with only the hCat mice

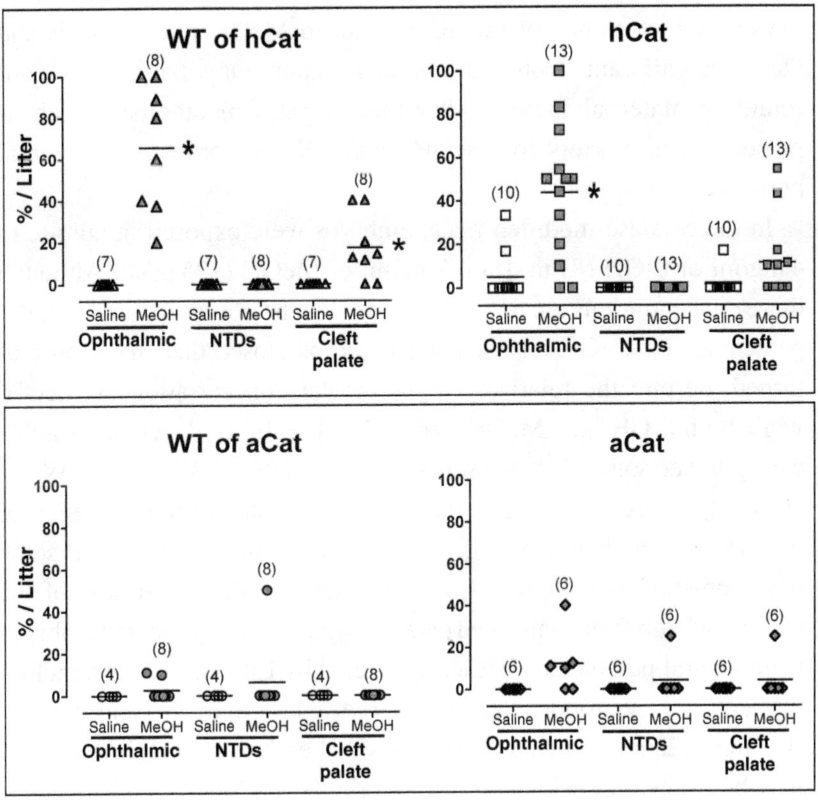

FIGURE 7.25 Effect of MeOH on embryopathies in mice with a genetic modulation of catalase *in vivo*. Pregnant dams were treated on gestational day (GD) 8 with two doses of 2 g/kg MeOH i.p. or its saline vehicle at 4-hour intervals, and assessed on GD 19. Each symbol represents one litter, with the number of litters shown in parentheses, and the mean by a horizontal bar. The symbol * indicates a difference from the respective saline treatment ($p < 0.01$ for ophthalmic anomalies and $p < 0.05$ for cleft palate anomalies). *Source*: From Siu et al. (2013).

showing a small increase in MeOH peak plasma concentration and AUC (Sweeting et al., 2010). The resistance of the C3H mice *in vivo* is consistent with the lower susceptibility of this strain to MeOH embryopathies in embryo culture compared to C57BL/6J embryos. In the hCat mice, it is possible that the 1.5-fold increase in embryonic catalase activity was insufficient to protect against MeOH-initiated ROS formation, although the same hCat embryos were protected in embryo culture. In light of the protective role for embryonic catalase evident in embryo culture using these mice, the *in vivo* resistance of the aCat strain to MeOH teratogenesis and the nonsignificant protection in hCat mice may be due to confounding maternal factors other than hepatic metabolism, such as placental transporters for MeOH and/or FA, or protective maternal hormones.

In the catalase-modified mice, embryos were exposed in culture to 4 mg/ml of EtOH (87 mM) or 4 mg/ml of MeOH (125 mM). Although these concentrations are not equimolar, it can be observed that a smaller percentage of MeOH-exposed WT embryos closed their neuropore or turned, despite the relatively higher molar concentration of MeOH, while both EtOH- and MeOH-exposed WT embryos developed significantly fewer somites than saline-exposed controls (Miller and Wells, 2011). These results suggest that the embryopathic molar potency of MeOH is less than that for EtOH, and substantially so for some developmental outcomes. Also interesting was that regardless of the molar concentration, enhanced catalase significantly protected embryos from several parameters that were affected by EtOH or MeOH, including anterior neuropore closure, somite development, and turning (Figures 7.23 and 7.24). This suggests that embryonic catalase activity may be a potential risk factor for both EtOH and MeOH, and enhanced embryonic catalase is protective.

7.4.3.2 Free Radical Spin Trapping Agent In embryo culture, MeOH-exposed CD-1 mouse embryos pretreated with the free radical spin trapping agent PBN exhibited reduced MeOH embryopathies, evidenced by increased anterior neuropore closure, turning, and somite development, suggesting ROS involvement in the mechanism of MeOH embryopathies (Miller and Wells, unpublished). This interpretation is consistent with the protective effect of PBN in reducing embryonic/fetal DNA oxidation and the embryopathic effects of other ROS-initiating teratogens in embryo culture and/or *in vivo* (Wells et al., 2009b; Lee et al., 2011).

7.4.4 Carcinogenic Potential

7.4.4.1 Oxidatively Damaged DNA *In vitro* and *in vivo* genotoxicity tests indicate MeOH is not mutagenic (IPCS, 1997), but carcinogenic potential has been claimed in one controversial long-term rodent cancer bioassay that has not been replicated (Soffritti et al., 2002). To determine if MeOH could indirectly damage DNA via ROS-mediated mechanisms, we treated male CD-1 mice, NZW rabbits, and cynomolgous monkeys with MeOH (2.0 g/kg i.p.) and 6 hours later assessed oxidative damage to DNA, measured as 8-oxodG formations by HPLC with electrochemical detection. We found no MeOH-dependent increases in 8-oxodG in bone marrow, spleen, lung, liver, or kidney of any species (Figure 7.26) (McCallum et al., 2011a,b). Chronic treatment of CD-1 mice with MeOH (2.0 g/kg i.p.) daily for 15 days also did not increase 8-oxodG levels in these organs. To rule out the possibility that the lack of effect of MeOH exposure on accumulation of oxidatively damaged DNA was due to masking by rapid repair of induced lesions, we performed further studies in DNA repair-deficient *oxoguanine glycosylase 1* (*Ogg1*) KO mice. The results were corroborated in untreated *Ogg1* KO mice, which accumulated 8-oxodG in bone marrow, spleen, lung, kidney, and liver with age, but exhibited no increase following MeOH treatment, despite a twofold increase in renal 8-oxodG in *Ogg1* KO mice following treatment with a ROS-initiating positive control, the renal carcinogen potassium bromate ($KBrO_3$; 100 mg/kg i.p.) (Figure 7.27) (McCallum et al., 2011a,b). These observations suggest that MeOH exposure does not promote the accumulation of oxidatively damaged DNA in lung, kidney, or liver, and that environmental exposure to MeOH is unlikely to initiate carcinogenesis in these organs by ROS-initiated DNA oxidation.

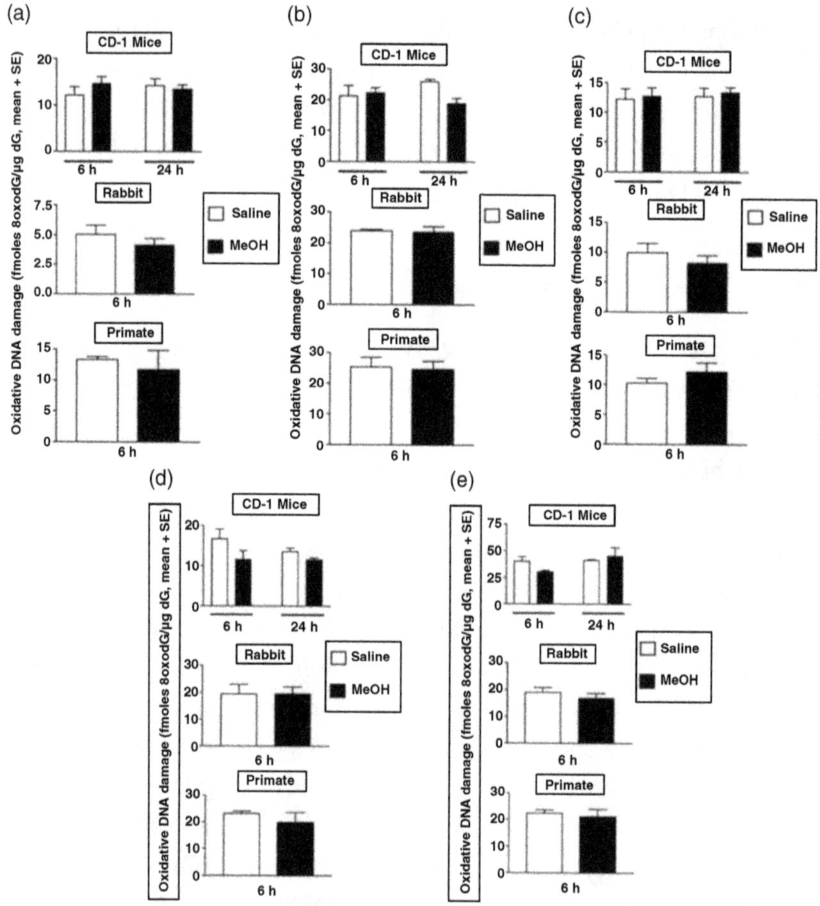

FIGURE 7.26 MeOH-initiated 8-oxo-2′deoxyguanine (8-oxodG) formation in the lung, liver, kidney, bone marrow, and spleen of male CD-1 mice, NZW rabbits, and cynomolgous monkeys in 6 hours. Animals were treated intraperitoneally with a single dose of 2.0 g/kg bw MeOH (20% [w/v] in sterile saline) or saline vehicle (controls) and sacrificed at 6 or 24 hours post-injection. Genomic DNA was isolated and analyzed for oxidatively damaged DNA damage reflected by the formation of 8-oxodG in the (a) lung, (b) liver, (c) kidney, (d) bone marrow, and (e) spleen of each species. Values are mean + SE; $N = 4$ for mice and $N = 3$ for rabbit and monkeys, respectively. *Source*: Modified from McCallum et al. (2011a,b).

7.4.4.2 Hydroxynonenal-Histidine Protein Adducts On the basis of our negative findings for MeOH-dependent ROS-mediated DNA damage and conflicting reports in the literature of MeOH-induced lipid peroxidation (Skrzydlewska et al., 2000; Parthasarathy et al., 2006a;

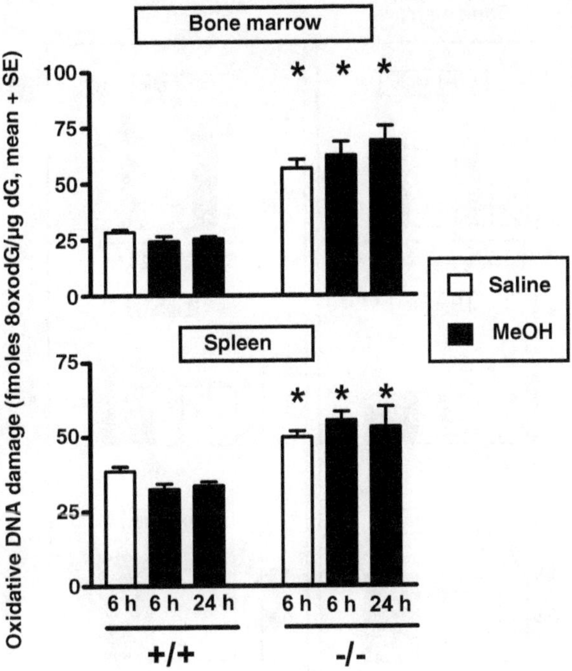

FIGURE 7.27 MeOH exposure does not increase 8-oxodG levels in bone marrow and spleen of *Ogg1* (+/+) or *Ogg1* (−/−) mice. Mice were given a single i.p. dose of 2.0 g/kg bw MeOH (20% [w/v] in sterile saline) or saline vehicle control, and sacrificed at 6 and 24 hours post-injection. Genomic DNA was isolated and analyzed for oxidative DNA damage reflected by the formation of 8-oxodG. Values are means + SE; $N = 4$. Symbol * denotes a difference in *Ogg1* (−/−) sample compared to the respective group in *Ogg1* (+/+) mice ($p < 0.05$). *Source*: Modified from McCallum et al. (2011a,b).

Parthasarathy et al., 2006b) determined by the controversial TBARS method (see Section 7.4.1.3), we conducted an additional measure of free radical-initiated macromolecular damage by measuring the HNE-histidine (HNE-His) protein adducts under conditions minimizing the influence of adventitious iron in cellular homogenates (see Section 7.4.1.3 for carcinogenic relevance). We detected modest increases in the levels of the HNE-His adduct only in mouse bone marrow (1.4-fold) and in rabbit spleen (1.5-fold) with no increases observed in primate bone marrow or spleen or mouse spleen (McCallum et al., 2011a) (Figure 7.28). For comparison, acute exposure to 20 mg/kg i.p. of the

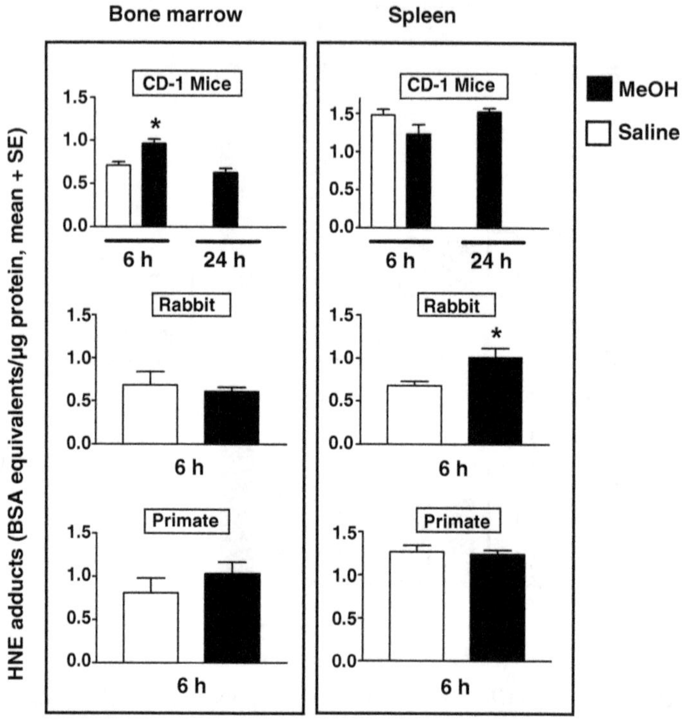

FIGURE 7.28 HNE-His protein adducts in bone marrow and spleen from male CD-1 mice, New Zealand white rabbits, or cynomolgus monkeys following acute exposure to MeOH (2.0 g/kg i.p.). Animals were given a single i.p. dose of 2.0 g/kg bw MeOH (20% [w/v] in sterile saline) or saline vehicle (control) and sacrificed at 6 or 24 hours post-injection. Values are means + SE; $N = 4$ for mice and $N = 3$ for rabbits and monkeys. The symbol * denotes a difference from saline control ($p < 0.05$). *Source*: Modified from McCallum et al. (2011a).

redox cycling agent doxorubicin increased the levels of HNE-His protein adducts by 5.3-fold in mouse heart homogenates (Mukhopadhyay et al., 2010). In contrast, the absence of MeOH-initiated HNE-His protein adducts in mouse spleen, and more importantly in either primate bone marrow or spleen, or in rabbit bone marrow, indicates this oxidative effect of MeOH is limited and unlikely to occur to a biologically significant degree in adult humans. In general, these observations suggest that a limit dose of MeOH (2.0 g/kg i.p.) can produce modest oxidative stress, but the potential for macromolecular damage is highly specific with respect to macromolecular target, species, and tissue. Moreover, perhaps owing to a combination of subcellular localization, packaging in nucleosomes, and repair mechanisms, DNA is more resistant than lipids to MeOH-initiated oxidative damage in adults.

7.5 CONCLUSIONS

Our results suggest the following:

- As the metabolism of MeOH by rabbits is more similar than mice to that in humans, the rabbit might be a more accurate model than the mouse, and perhaps the rat, for predicting the human risk for MeOH developmental toxicity.
- Given that rabbits, which more closely reflect human MeOH metabolism, and at least one strain of mice and one strain of rat, are resistant to MeOH teratogenesis, it is questionable whether the human risk for MeOH developmental toxicity can be accurately assessed in sensitive rodent models.
- The respectively enhanced and reduced susceptibility of aCat and hCat mice to the embryopathic effects of MeOH in embryo culture, albeit not *in vivo*, suggest that ROS may contribute to the underlying mechanism of MeOH teratogenicity in rodents. These results also suggest that embryonic catalase plays an important protective role via its antioxidative activity, as distinct from its peroxidative, MeOH-metabolizing role in maternal liver or the embryo. The absence of modulation by altered catalase expression *in vivo* merits further investigation.
- The reduced embryopathies observed in MeOH-exposed mouse embryos pretreated with a free radical spin trapping agent in whole embryo culture further support a role of oxidative stress in the mechanism of MeOH developmental toxicity.
- MeOH appears to be less embryopathic than EtOH on a molar basis, based on mouse embryo culture results, although it is difficult to extrapolate this comparison to *in vivo* studies in susceptible rodent strains where conditions are highly variable.

- In regard to the mechanism by which MeOH enhances embryopathic ROS formation, our preliminary results showing a reduction in MeOH embryopathies in culture by pretreatment with a NOX inhibitor suggest that MeOH may activate/induce embryonic NOXs that form ROS intracellularly. Conversely, PHS-mediated bioactivation of MeOH to a free radical intermediate is unlikely to play a role, as preliminary studies using PHS inhibition did not alter MeOH-initiated embryopathies in culture.
- MeOH did not increase the level of oxidatively damaged DNA in any tissue in any species, despite enhanced baseline DNA oxidation observed in DNA repair-deficient *Ogg1* KO mice, and enhanced DNA oxidation observed with a ROS-initiating positive control. Similarly, levels of HNE-His protein adducts, reflecting free radical-mediated production of the potentially carcinogenic lipid peroxidation product HNE, were not enhanced by MeOH in primate bone marrow or spleen, or in rabbit bone marrow or mouse spleen, although modest increases were observed in rabbit spleen and mouse bone marrow. These results suggest that it is unlikely that human environmental exposure to MeOH would cause cancer via a mechanism involving oxidatively damaged DNA.

ACKNOWLEDGMENT

These studies were supported by grants from the Methanol Foundation (USA) and the Canadian Institutes of Health Research.

REFERENCES

Abel, E.L. (1982) In utero alcohol exposure and developmental delay of response inhibition. *Alcohol. Clin. Exp. Res.* **6**, 369–376.

Abramov, J.P. and Wells, P.G. (2011) Embryonic catalase protects against endogenous and phenytoin-enhanced DNA oxidation and embryopathies in acatalasemic and human catalase-expressing mice. *FASEB J.* **25**, 2188–2200.

AICR (2007) *Food, Nutrition, Physical Activity, and the Prevention of Cancer: a Global Perspective*. World Cancer Research Fund/American Institute for Cancer Research, Washington DC.

Algar, E.M., Seeley, T.L., and Holmes, R.S. (1983) Purification and molecular properties of mouse alcohol dehydrogenase isozymes. *Eur. J. Biochem./FEBS* **137**, 139–147.

Ames, B.N. (1989) Mutagenesis and carcinogenesis: endogenous and exogenous factors. *Environ. Mol. Mutagen.* **14**, (Suppl 16), 66–77.

Andrews, J.E., Ebron-McCoy, M., Logsdon, T.R., Mole, L.M., Kavlock, R.J., and Rogers, J.M. (1993) Developmental toxicity of methanol in whole embryo culture: a comparative study with mouse and rat embryos. *Toxicology* **81**, 205–215.

Apaja, M. (1980) Evaluation of toxicity and carcinogenicity of malonaldehyde. *Acta Universitatis Ouluensis* **55**, 28–59.

Baan, R., Straif, K., Grosse, Y., Secretan, B., El Ghissassi, F., Bouvard, V., Altieri, A., and Cogliano, V. (2007) Carcinogenicity of alcoholic beverages. *Lancet Oncol.* **8**, 292–293.

Baeuerle, P.A. and Baltimore, D. (1996) NF-kappa B: ten years after. *Cell* **87**, 13–20.

Baumbach, G.L., Cancilla, P.A., Martin-Amat, G., Tephly, T.R., McMartin, K.E., Makar, A.B., Hayreh, M.S., and Hayreh, S.S. (1977) Methyl alcohol poisoning. IV. Alterations of the morphological findings of the retina and optic nerve. *Arch. Ophthalmol.* **95**, 1859–1865.

Becker, H.C. and Randall, C.L. (1989) Effects of prenatal ethanol exposure in C57BL mice on locomotor activity and passive avoidance behavior. *Psychopharmacology* **97**, 40–44.

Becker, H.C., Diaz-Granados, J.L., and Randall, C.L. (1996) Teratogenic actions of ethanol in the mouse: a minireview. *Pharmacol. Biochem. Behav.* **55**, 501–513.

Belson, M. and Morgan, B.W. (2004) Methanol toxicity in a newborn. *J. Toxicol. Clin. Toxicol.* **42**(5), 673–677.

Berlett, B.S. and Stadtman, E.R. (1997) Protein oxidation in aging, disease, and oxidative stress. *J. Biol. Chem.* **272**, 20313–20316.

Black, K.A., Eells, J.T., Noker, P.E., Hawtrey, C.A., and Tephly, T.R. (1985) Role of hepatic tetrahydrofolate in the species difference in methanol toxicity. *Proc. Natl. Acad. Sci. U S A.* **82**, 3854–3858.

Blomstrand, R. and Ingemansson, S.O. (1984) Studies on the effect of 4-methylpyrazole on methanol poisoning using the monkey as an animal model: with particular reference to the ocular toxicity. *Drug Alcohol Depend.* **13**, 343–355.

Bolon, B., Dorman, D.C., Janszen, D., Morgan, K.T., and Welsch, F. (1993) Phase-specific developmental toxicity in mice following maternal methanol inhalation. *Fundam. Appl. Toxicol.* **21**, 508–516.

Bolon, B., Welsch, F., and Morgan, K.T. (1994) Methanol-induced neural tube defects in mice: pathogenesis during neurulation. *Teratology* **49**, 497–517.

Bradford, B.U., Seed, C.B., Handler, J.A., Forman, D.T., and Thurman, R.G. (1993) Evidence that catalase is a major pathway of ethanol oxidation in vivo: dose-response studies in deer mice using methanol as a selective substrate. *Arch. Biochem. Biophys.* **303**, 172–176.

Breinholt, V., Lauridsen, S.T., Daneshvar, B., and Jakobsen, J. (2000) Dose-response effects of lycopene on selected drug-metabolizing and antioxidant enzymes in the rat. *Cancer Lett.* **154**, 201–210.

Brent, J. (2009) Fomepizole for ethylene glycol and methanol poisoning. *N. Engl. J. Med.* **360**, 2216–2223.

Brooks, P.J., Enoch, M.A., Goldman, D., Li, T.K., and Yokoyama, A. (2009) The alcohol flushing response: an unrecognized risk factor for esophageal cancer from alcohol consumption. *PLoS Med.* **6**, e50.

Brown, D.I. and Griendling, K.K. (2009) Nox proteins in signal transduction. *Free Radic. Biol. Med.* **47**, 1239–1253.

Buhler, R., Hempel, J., Kaiser, R., von Wartburg, J.P., Vallee, B.L., and Jornvall, H. (1984) Human alcohol dehydrogenase: structural differences between the beta and gamma subunits suggest parallel duplications in isoenzyme evolution and predominant expression of separate gene descendants in livers of different mammals. *Proc. Natl. Acad. Sci. USA.* **81**, 6320–6324.

Burbacher, T., Shen, D., Grant, K., Sheppard, L., Damian, D., Ellis, S., and Liberato, N. (1999) Reproductive and offspring developmental effects following maternal inhalation exposure to methanol in nonhuman primates. *Res. Rep. Health Eff. Inst.* **i–ii**, 1–117; discussion 119–133.

Castro, G.D., Costantini, M.H., Delgado de Layno, A.M., and Castro, J.A. (2002) Rat liver microsomal and nuclear activation of methanol to hydroxymethyl free radicals. *Toxicol. Lett.* **129**, 227–236.

Cederbaum, A.I. and Qureshi, A. (1982) Role of catalase and hydroxyl radicals in the oxidation of methanol by rat liver microsomes. *Biochem. Pharmacol.* **31**, 329–335.

Chernoff, G.F. (1980) The fetal alcohol syndrome in mice: maternal variables. *Teratology* **22**, 71–75.

Choudhury, S., Pan, J., Amin, S., Chung, F.L., and Roy, R. (2004) Repair kinetics of trans-4-hydroxynonenal-induced cyclic 1,N2-propanodeoxyguanine DNA adducts by human cell nuclear extracts. *Biochemistry* **43**, 7514–7521.

Christmann, M., Tomicic, M.T., Roos, W.P., and Kaina, B. (2003) Mechanisms of human DNA repair: an update. *Toxicology* **193**, 3–34.

Clary, J.J. (2003) Methanol, is it a developmental risk to humans? *Regul. Toxicol. Pharmacol.* **37**, 83–91.

Cook, R.J., Champion, K.M., and Giometti, C.S. (2001) Methanol toxicity and formate oxidation in NEUT2 mice. *Arch. Biochem. Biophys.* **393**, 192–198.

Cruzan, G. (2009) Assessment of the cancer potential of methanol. *Crit. Rev. Toxicol.* **39**, 347–363.

Cummings, A.M. (1993) Evaluation of the effects of methanol during early pregnancy in the rat. *Toxicology* **79**, 205–214.

De-Carvalho, R.R., Delgado, I.F., Souza, C.A., Chahoud, I., and Paumgartten, F.J. (1994) Embryotoxicity of methanol in well-nourished and malnourished rats. *Braz. J. Med. Biol. Res. = Revista brasileira de pesquisas medicas e biologicas/Sociedade Brasileira de Biofisica . . . [et al.]* **27**, 2915–2923.

Degitz, S.J., Rogers, J.M., Zucker, R.M., and Hunter, E.S., 3rd (2004) Developmental toxicity of methanol: pathogenesis in CD-1 and C57BL/6J mice exposed in whole embryo culture. *Birth Defects Res. A Clin. Mol. Teratol.* **70**, 179–184.

Dobrzynska, I., Skrzydlewska, E., and Figaszewski, Z. (1999) Influence of trolox derivative and N-acetylcysteine on surface charge density of erythrocytes in methanol intoxicated rats. *Environ. Toxicol. Pharmacol.* **8**, 15–21.

Dobrzynska, I., Skrzydlewska, E., Kasacka, I., and Figaszewski, Z. (2000) Protective effect of N-acetylcysteine on rat liver cell membrane during methanol intoxication. *J. Pharm. Pharmacol.* **52**, 547–552.

Dong, J., Sulik, K.K., and Chen, S.Y. (2010) The role of NOX enzymes in ethanol-induced oxidative stress and apoptosis in mouse embryos. *Toxicol. Lett.* **193**, 94–100.

Ekins, B.R., Rollins, D.E., Duffy, D.P., and Gregory, M.C. (1985) Standardized treatment of severe methanol poisoning with ethanol and hemodialysis. *West. J. Med.* **142**, 337–340.

Esterbauer, H., Schaur, R.J., and Zollner, H. (1991) Chemistry and biochemistry of 4-hydroxynonenal, malonaldehyde and related aldehydes. *Free Radic. Biol. Med.* **11**, 81–128.

Evans, M.D., Dizdaroglu, M., and Cooke, M.S. (2004) Oxidative DNA damage and disease: induction, repair and significance. *Mutat. Res.* **567**, 1–61.

Fam, S.S. and Morrow, J.D. (2003) The isoprostanes: unique products of arachidonic acid oxidation-a review. *Curr. Med. Chem.* **10**, 1723–1740.

Feng, Z., Hu, W., and Tang, M.S. (2004) Trans-4-hydroxy-2-nonenal inhibits nucleotide excision repair in human cells: a possible mechanism for lipid peroxidation-induced carcinogenesis. *PNAS* **101**, 8598–8602.

Fukui, Y., Naruse, I., Hoshino, K., Kameyama, Y., and Takeuchi, Y. (1990) Strain difference in effects of ethanol treatment on fetal development in C57Bl and DBA mice. *Congenital Anomalies* **30**, 177–186.

Gilliam, D.M., Stilman, A., Dudek, B.C., and Riley, E.P. (1987) Fetal alcohol effects in long- and short-sleep mice: activity, passive avoidance, and in utero ethanol levels. *Neurotoxicol. Teratol.* **9**, 349–357.

Goyal, M.M. and Basak, A. (2010) Human catalase: looking for complete identity. *Protein & Cell* **1**, 888–897.

Gutteridge, J.M. (1984) Lipid peroxidation initiated by superoxide-dependent hydroxyl radicals using complexed iron and hydrogen peroxide. *FEBS Lett.* **172**, 245–249.

Gutteridge, J.M. and Quinlan, G.J. (1983) Malondialdehyde formation from lipid peroxides in the thiobarbituric acid test: the role of lipid radicals, iron salts, and metal chelators. *J. Appl. Biochem.* **5**, 293–299.

Haffner, H.T., Wehner, H.D., Scheytt, K.D., and Besserer, K. (1992) The elimination kinetics of methanol and the influence of ethanol. *Int. J. Legal Med.* **105**, 111–114.

Halliwell, B. and Gutteridge, J.M. (2007) *Free Radicals in Biology and Medicine*. Oxford University Press, Oxford.

Hansen, J.M., Contreras, K.M., and Harris, C. (2005) Methanol, formaldehyde, and sodium formate exposure in rat and mouse conceptuses: a potential role of the visceral yolk sac in embryotoxicity. *Birth Defects Res. A Clin. Mol. Teratol.* **73**, 72–82.

Hantson, P., Lambermont, J.Y., and Mahieu, P. (1997) Methanol poisoning during late pregnancy. *J. Toxicol. Clin. Toxicol.* **35**(2), 187–191.

Harris, C., Dixon, M., and Hansen, J.M. (2004) Glutathione depletion modulates methanol, formaldehyde and formate toxicity in cultured rat conceptuses. *Cell Biol. Toxicol.* **20**, 133–145.

Harris, C., Wang, S.W., Lauchu, J.J., and Hansen, J.M. (2003) Methanol metabolism and embryotoxicity in rat and mouse conceptuses: comparisons of alcohol dehydrogenase (ADH1), formaldehyde dehydrogenase (ADH3), and catalase. *Reprod. Toxicol. (Elmsford, NY)* **17**, 349–357.

Hayreh, M.S., Hayreh, S.S., Baumbach, G.L., Cancilla, P., Martin-Amat, G., Tephly, T.R., McMartin, K.E., and Makar, A.B. (1977) Methyl alcohol poisoning III. Ocular toxicity. *Arch. Ophthalmol.* **95**, 1851–1858.

Hines, R.N. (2008) The ontogeny of drug metabolism enzymes and implications for adverse drug events. *Pharmacol. Ther.* **118**, 250–267.

Horton, V.L., Higuchi, M.A., and Rickert, D.E. (1992) Physiologically based pharmacokinetic model for methanol in rats, monkeys, and humans. *Toxicol. Appl. Pharmacol.* **117**, 26–36.

Hovda, K.E. and Jacobsen, D. (2008) Expert opinion: fomepizole may ameliorate the need for hemodialysis in methanol poisoning. *Hum. Exp. Toxicol.* **27**, 539–546.

Hovda, K.E., Mundal, H., Urdal, P., McMartin, K., and Jacobsen, D. (2007) Extremely slow formate elimination in severe methanol poisoning: a fatal case report. *Clin. Toxicol. (Philadelphia, Pa)* **45**, 516–521.

Hsu, G.W., Ober, M., Carell, T., and Beese, L.S. (2004) Error-prone replication of oxidatively damaged DNA by a high-fidelity DNA polymerase. *Nature.* **431**, 217–221.

IARC (2006) IARC monographs on the evaluation of carcinogenic risks to humans volume 88: formaldehyde, 2-butoxyethanol and 1-tert-butoxypropan-2-ol. World Health Organization, Geneva.

Infurna, R. and Weiss, B. (1986) Neonatal behavioral toxicity in rats following prenatal exposure to methanol. *Teratology.* **33**, 259–265.

IPCS (1997) Environmental Health Criteria 196 - Methanol. World Health Organization, Geneva.

Ishii, K., Zhen, L.X., Wang, D.H., Funamori, Y., Ogawa, K., and Taketa, K. (1996) Prevention of mammary tumorigenesis in acatalasemic mice by vitamin E supplementation. *Jpn. J. Cancer Res.* **87**, 680–684.

Janssen-Heininger, Y.M., Mossman, B.T., Heintz, N.H., Forman, H.J., Kalyanaraman, B., Finkel, T., Stamler, J.S., Rhee, S.G., and van der Vliet, A. (2008) Redox-based regulation of signal transduction: principles, pitfalls, and promises. *Free Radic. Biol. Med.* **45**, 1–17.

Jiang, F., Zhang, Y., and Dusting, G.J. (2011) NADPH oxidase-mediated redox signaling: roles in cellular stress response, stress tolerance, and tissue repair. *Pharmacol. Rev.* **63**, 218–242.

Johlin, F.C., Fortman, C.S., Nghiem, D.D., and Tephly, T.R. (1987) Studies on the role of folic acid and folate-dependent enzymes in human methanol poisoning. *Mol. Pharmacol.* **31**, 557–561.

Jones, K.L. (2011) The effects of alcohol on fetal development. *Birth Defects Res. C Embryo Today* **93**, 3–11.

Jones, K.L. and Smith, D.W. (1973) Recognition of the fetal alcohol syndrome in early infancy. *Lancet.* **302**, 999–1001.

Juchau, M.R., Boutelet-Bochan, H., and Huang, Y. (1998) Cytochrome-P450-dependent biotransformation of xenobiotics in human and rodent embryonic tissues. *Drug Metab. Rev.* **30**, 541–568.

Kadiiska, M.B. and Mason, R.P. (2000) Acute methanol intoxication generates free radicals in rats: an ESR spin trapping investigation. *Free Radic. Biol. Med.* **28**, 1106–1114.

Karinje, K.U. and Ogata, M. (1990) Methanol metabolism in acatalasemic mice. *Physiol. Chem. Phys. Med. NMR.* **22**, 193–198.

Kavet, R. and Nauss, K.M. (1990) The toxicity of inhaled methanol vapors. *Crit. Rev. Toxicol.* **21**, 21–50.

Kedishvili, N.Y., Bosron, W.F., Stone, C.L., Hurley, T.D., Peggs, C.F., Thomasson, H.R., Popov, K.M., Carr, L.G., Edenberg, H.J., and Li, T.K. (1995) Expression and kinetic characterization of recombinant human stomach alcohol dehydrogenase. Active-site amino acid sequence explains substrate specificity compared with liver isozymes. *J. Biol. Chem.* **270**, 3625–3630.

Kennedy, J.C., Memet, S., and Wells, P.G. (2004) Antisense evidence for nuclear factor-kappa B-dependent embryopathies initiated by phenytoin-enhanced oxidative stress. *Mol. Pharmacol.* **66**, 404–412.

Kerns, W.D., Pavkov, K.L., Donofrio, D.J., Gralla, E.J., and Swenberg, J.A. (1983) Carcinogenicity of formaldehyde in rats and mice after long-term inhalation exposure. *Cancer Res.* **43**, 4382–4392.

Kinoshita, A., Wanibuchi, H., Morimura, K., Wei, M., Nakae, D., Arai, T., Minowa, O., Noda, T., Nishimura, S., and Fukushima, S. (2007) Carcinogenicity of dimethylarsinic acid in Ogg1-deficient mice. *Cancer Science* **98**, 803–814.

Kirkman, H.N. and Gaetani, G.F. (1984) Catalase: a tetrameric enzyme with four tightly bound molecules of NADPH. *Proc. Natl. Acad. Sci. USA.* **81**, 4343–4347.

Klaassen, C.D. (2008) *Cassarett and Doull's Toxicology: The Basic Science of Poisons.* McGraw-Hill Companies, Inc., New York.

Klaunig, J.E. and Kamendulis, L.M. (2004) The role of oxidative stress in carcinogenesis. *Annu. Rev. Pharmacol. Toxicol.* **44**, 239–267.

Knobloch, J. and Ruther, U. (2008) Shedding light on an old mystery: thalidomide suppresses survival pathways to induce limb defects. *Cell Cycle* **7**, 1121–1127.

Knobloch, J., Schmitz, I., Gotz, K., Schulze-Osthoff, K., and Ruther, U. (2008) Thalidomide induces limb anomalies by PTEN stabilization, Akt suppression, and stimulation of caspase-dependent cell death. *Mol. Cell. Biol.* **28**, 529–538.

Knobloch, J., Shaughnessy, J.D., Jr., and Ruther, U. (2007) Thalidomide induces limb deformities by perturbing the Bmp/Dkk1/Wnt signaling pathway. *FASEB J* **21**, 1410–1421.

Koop, D.R. (2006) Alcohol metabolism's damaging effects on the cell: a focus on reactive oxygen generation by the enzyme cytochrome P450 2E1. *Alcohol Res. Health* **29**, 274–280.

Kotch, L.E. and Sulik, K.K. (1992) Experimental fetal alcohol syndrome: proposed pathogenic basis for a variety of associated facial and brain anomalies. *Am. J. Med. Genet.* **44**, 168–176.

Kunitoh, S., Tanaka, T., Imaoka, S., Funae, Y., and Monna, Y. (1993) Contribution of cytochrome P450s to MEOS (microsomal ethanol-oxidizing system): a specific and sensitive assay of MEOS activity by HPLC with fluorescence labeling. *Alcohol and Alcohol. (Oxford, Oxfordshire)* **1B**, 63–68.

Lambeth, J.D. (2004) NOX enzymes and the biology of reactive oxygen. *Nat. Rev. Immunol.* **4**, 181–189.

Lander, H.M. (1997) An essential role for free radicals and derived species in signal transduction. *FASEB J.* **11**, 118–124.

Lee, C.J., Goncalves, L.L., and Wells, P.G. (2011) Embryopathic effects of thalidomide and its hydrolysis products in rabbit embryo culture: evidence for a prostaglandin H synthase (PHS)-dependent, reactive oxygen species (ROS)-mediated mechanism. *FASEB J.* **25**, 2468–2483.

Levine, R.L., Williams, J.A., Stadtman, E.R., and Shacter, E. (1994) Carbonyl assays for determination of oxidatively modified proteins. *Methods Enzymol* **233**, 346–357.

Lieber, C.S. (2004) The discovery of the microsomal ethanol oxidizing system and its physiologic and pathologic role. *Drug Metab. Rev.* **36**, 511–529.

Lu, K., Collins, L.B., Ru, H., Bermudez, E., and Swenberg, J.A. (2010) Distribution of DNA adducts caused by inhaled formaldehyde is consistent with induction of nasal carcinoma but not leukemia. *Toxicol. Sci.* **116**, 441–451.

Lutwak-Mann, C. (1938) Alcohol dehydrogenase of animal tissues. *Biochem. J.* **32**, 1364–1374.

Makar, A.B. and Tephly, T.R. (1977) Methanol poisoning VI: role of folic acid in the production of methanol poisoning in the rat. *J. Toxicol. Environ. Health* **2**, 1201–1209.

Makar, A.B., Tephly, T.R., and Mannering, G.J. (1968) Methanol metabolism in the monkey. *Mol. Pharmacol.* **4**, 471–483.

Mambo, E., Chatterjee, A., de Souza-Pinto, N.C., Mayard, S., Hogue, B.A., Hoque, M.O., Dizdaroglu, M., Bohr, V.A., and Sidransky, D. (2005) Oxidized guanine lesions and hOgg1 activity in lung cancer. *Oncogene*. **24**, 4496–4508.

Mani, J.C., Pietruszko, R., and Theorell, H. (1970) Methanol activity o alcohol dehydrogenases from human liver, horse liver, and yeast. *Arch. Biochem. Biophys.* **140**, 52–59.

Mannering, G.J., Van Harken, D.R., Makar, A.B., Tephly, T.R., Watkins, W.D., and Goodman, J.I. (1969) Role of the intracellular distribution of hepatic catalase in the peroxidative oxidation of methanol. *Ann. NY. Acad. Sci.* **168**, 265–280.

Maritim, A.C., Sanders, R.A., and Watkins, J.B., 3rd (2003) Diabetes, oxidative stress, and antioxidants: a review. *J. Biochem. Mol. Toxicol.* **17**, 24–38.

Marnett, L.J. (2000) Oxyradicals and DNA damage. *Carcinogenesis*. **21**, 361–370.

Mattson, S.N., Schoenfeld, A.M., and Riley, E.P. (2001) Teratogenic effects of alcohol on brain and behavior. *Alcohol Res. Health* **25**, 185–191.

McCallum, G.P., Siu, M., Ondovcik, S.L., Sweeting, J.N., and Wells, P.G. (2011a) Methanol exposure does not lead to accumulation of oxidative DNA damage in bone marrow and spleen of mice, rabbits or primates. *Mol. Carcinog.* **50**, 163–172.

McCallum, G.P., Siu, M., Sweeting, J.N., and Wells, P.G. (2011b) Methanol exposure does not produce oxidatively damaged DNA in lung, liver or kidney of adult mice, rabbits or primates. *Toxicol. Appl. Pharmacol.* **250**, 147–153.

McMartin, K.E., Martin-Amat, G., Makar, A.B., and Tephly, T.R. (1977) Methanol poisoning. V. Role of formate metabolism in the monkey. *J. Pharmacol. Exp. Ther.* **201**, 564–572.

McMartin, K.E., Martin-Amat, G., Noker, P.E., and Tephly, T.R. (1979) Lack of a role for formaldehyde in methanol poisoning in the monkey. *Biochem. Pharmacol.* **28**, 645–649.

McMartin, K.E., Virayotha, V., and Tephly, T.R. (1981) High-pressure liquid chromatography separation and determination of rat liver folates. *Arch. Biochem. Biophys.* **209**, 127–136.

Miller, L. and Wells, P.G. (2011) Altered methanol embryopathies in embryo culture with mutant catalase-deficient mice and transgenic mice expressing human catalase. *Toxicol. Appl. Pharmacol.* **252**, 55–61.

Moeller, B.C., Lu, K., Doyle-Eisele, M., McDonald, J., Gigliotti, A., and Swenberg, J.A. (2011) Determination of N2-hydroxymethyl-dG adducts in the nasal epithelium and bone marrow of nonhuman primates following 13CD2-formaldehyde inhalation exposure. *Chem. Res. Toxicol.* **24**, 162–164.

Mukhopadhyay, P., Rajesh, M., Batkai, S., Patel, V., Kashiwaya, Y., Liaudet, L., Evgenov, O.V., Mackie, K., Hasko, G., and Pacher, P. (2010) CB1 cannabinoid receptors promote oxidative stress and cell death in murine models of doxorubicin-induced cardiomyopathy and in human cardiomyocytes. *Cardiovasc. Res.* **85**, 773–784.

NEDO (1985) Toxicological research of methanol as a fuel for power stations. New Energy Development Organization, Tokyo, Japan.

NEDO (1987) Toxicological research of methanol as a fuel for power stations: summary report on tests with monkeys, rats and mice. New Energy Development Organization, Tokyo, Japan.

Neeley, W.L. and Essigmann, J.M. (2006) Mechanisms of formation, genotoxicity, and mutation of guanine oxidation products. *Chem. Res. Toxicol.* **19**, 491–505.

Nelson, B.K., Brightwell, W.S., MacKenzie, D.R., Khan, A., Burg, J.R., Weigel, W.W., and Goad, P.T. (1985) Teratological assessment of methanol and ethanol at high inhalation levels in rats. *Fundam. Appl. Toxicol.* **5**, 727–736.

Nicholls, P. (1975) Formate as an inhibitor of cytochrome c oxidase. *Biochem. Biophys. Res. Commun.* **67**, 610–616.

Nicholls, P. (1976) The effect of formate on cytochrome aa3 and on electron transport in the intact respiratory chain. *Biochim. Biophys. Acta.* **430**, 13–29.

NTP (2002) NTP-CERHR expert panel report on the reproductive and developmental toxicity of methanol. In *National Toxicology Program.* U.S. Department of Health and Human Services.

NTP (2004) NTP-CERHR expert panel report on the reproductive and developmental toxicity of methanol. In *National Toxicology Program.* U.S. Department of Health and Human Services.

Nystrom, T. (2005) Role of oxidative carbonylation in protein quality control and senescence. *EMBO J.* **24**, 1311–1317.

Oshino, N., Jamieson, D., Sugano, T., and Chance, B. (1975) Optical measurement of the catalase-hydrogen peroxide intermediate (Compound I) in the liver of anaesthetized rats and its implication to hydrogen peroxide production in situ. *Biochem. J.* **146**, 67–77.

Osterloh, J.D., D'Alessandro, A., Chuwers, P., Mogadeddi, H., and Kelly, T.J. (1996) Serum concentrations of methanol after inhalation at 200 ppm. *Occup. Environ. Med. American College of Occupational and Environmental Medicine* **38**, 571–576.

Parman, T., Wiley, M.J., and Wells, P.G. (1999) Free radical-mediated oxidative DNA damage in the mechanism of thalidomide teratogenicity. *Nat. Med.* **5**, 582–585.

Parnell, S.E., O'Leary-Moore, S.K., Godin, E.A., Dehart, D.B., Johnson, B.W., Allan Johnson, G., Styner, M.A., and Sulik, K.K. (2009) Magnetic resonance microscopy defines ethanol-induced brain abnormalities in prenatal mice: effects of acute insult on gestational day 8. *Alcohol Clin. Exp. Res.* **33**, 1001–1011.

Parola, M., Bellomo, G., Robino, G., Barrera, G., and Dianzani, M.U. (1999) 4-Hydroxynonenal as a biological signal: molecular basis and pathophysiological implications. *Antioxid Redox Signal* **1**, 255–284.

Parthasarathy, N.J., Kumar, R.S., Manikandan, S., and Devi, R.S. (2006a) Methanol-induced oxidative stress in rat lymphoid organs. *J. Occup. Health* **48**, 20–27.

Parthasarathy, N.J., Kumar, R.S., Manikandan, S., Narayanan, G.S., Kumar, R.V., and Devi, R.S. (2006b) Effect of methanol-induced oxidative stress on the neuroimmune system of experimental rats. *Chem. Biol. Interact.* **161**, 14–25.

Paula, E.M., Mathangi, D.C., and Namasivayam, A. (2003) Free radical changes in methanol toxicity. *Indian J. Physiol. Pharmacol.* **47**, 207–211.

Paz-Elizur, T., Krupsky, M., Blumenstein, S., Elinger, D., Schechtman, E., and Livneh, Z. (2003) DNA repair activity for oxidative damage and risk of lung cancer. *J. Natl. Cancer Inst.* **95**, 1312–1319.

Peng, S.L., Fatenejad, S., and Craft, J. (1997) Scleroderma: a disease related to damaged proteins? *Nat. Med.* **3**, 276–278.

Perkins, R.A., Ward, K.W., and Pollack, G.M. (1995) A pharmacokinetic model of inhaled methanol in humans and comparison to methanol disposition in mice and rats. *Environ. Health Perspect.* **103**, 726–733.

Pietruszko, R. (1975a) Human liver alcohol dehydrogenase--inhibition of methanol activity by pyrazole, 4-methylpyrazole, 4-hydroxymethylpyrazole and 4-carboxypyrazole. *Biochem. Pharmacol.* **24**, 1603–1607.

Pietruszko, R. (1975b) Mammalian liver alcohol dehydrogenases. *Adv. Exp. Med. Biol.* **56**, 1–31.

Pikkarainen, P.H. and Raiha, N.C. (1967) Development of alcohol dehydrogenase activity in the human liver. *Pediatr. Res.* **1**, 165–168.

Poliakov, E., Meer, S.G., Roy, S.C., Mesaros, C., and Salomon, R.G. (2004) Iso [7]LGD2-protein adducts are abundant in vivo and free radical-induced oxidation of an arachidonyl phospholipid generates this D series isolevuglandin in vitro. *Chem. Res. Toxicol.* **17**, 613–622.

Randall, C.L., Taylor, J., and Walker, D.W. (1977) Ethanol-induced malformations in mice. *Alcohol Clin. Exp. Res.* **1**, 219–224.

Rang, H., Dale, M., Ritter, J., and Moore, P. (2003) *Pharmacology*, 5th edition. Churchill Livingstone, London.

Reddy, J.K., Lalwani, N.D., Qureshi, S.A., Reddy, M.K., and Moehle, C.M. (1984) Induction of hepatic peroxisome proliferation in nonrodent species, including primates. *Am. J. Pathol.* **114**, 171–183.

Roberts, L.J.2nd and Fessel, J.P. (2004) The biochemistry of the isoprostane, neuroprostane, and isofuran pathways of lipid peroxidation. *Chem. Phys. Lipids* **128**, 173–186.

Roe, O. (1955) The metabolism and toxicity of methanol. *Pharmacol. Rev.* **7**, 399–412.

Rogers, J.M., Brannen, K.C., Barbee, B.D., Zucker, R.M., and Degitz, S.J. (2004) Methanol exposure during gastrulation causes holoprosencephaly, facial dysgenesis, and cervical vertebral malformations in C57BL/6J mice. *Birth Defects Res. B Dev. Reprod. Toxicol.* **71**, 80–88.

Rogers, J.M. and Mole, M.L. (1997) Critical periods of sensitivity to the developmental toxicity of inhaled methanol in the CD-1 mouse. *Teratology* **55**, 364–372.

Rogers, J.M., Mole, M.L., Chernoff, N., Barbee, B.D., Turner, C.I., Logsdon, T.R., and Kavlock, R.J. (1993) The developmental toxicity of inhaled methanol in the CD-1 mouse, with quantitative dose-response modeling for estimation of benchmark doses. *Teratology* **47**, 175–188.

Russo, M.T., De Luca, G., Degan, P., Parlanti, E., Dogliotti, E., Barnes, D.E., Lindahl, T., Yang, H., Miller, J.H., and Bignami, M. (2004) Accumulation of the oxidative base lesion 8-hydroxyguanine in DNA of tumor-prone mice defective in both the Myh and Ogg1 DNA glycosylases. *Cancer Res.* **64**, 4411–4414.

Saker, M., Soulimane Mokhtari, N., Merzouk, S.A., Merzouk, H., Belarbi, B., and Narce, M. (2008) Oxidant and antioxidant status in mothers and their newborns according to birthweight. *Eur. J. Obstet. Gynecol. Reprod. Biol.* **141**, 95–99.

Schaur, R.J. (2003) Basic aspects of the biochemical reactivity of 4-hydroxynonenal. *Mol. Aspects Med.* **24**, 149–159.

Shaham, J., Bomstein, Y., Meltzer, A., Kaufman, Z., Palma, E., and Ribak, J. (1996) DNA–protein crosslinks, a biomarker of exposure to formaldehyde–in vitro and in vivo studies. *Carcinogenesis* **17**, 121–125.

Shimizu, T., Kondo, K., and Hayaishi, O. (1981) Role of prostaglandin endoperoxides in the serum thiobarbituric acid reaction. *Arch. Biochem. Biophys.* **206**, 271–276.

Siu, M. , Wiley, M.J., and Wells, P.G. (2013) Methanol teratogenicity in mutant mice with deficient catalase activity and transgenic mice expressing human catalase. Reprod. Toxicol. (in press).

Skrzydlewska, E., Elas, M., Farbiszewski, R., and Roszkowska, A. (2000) Effect of methanol intoxication on free-radical induced protein oxidation. *J. Appl. Toxicol.* **20**, 239–243.

Skrzydlewska, E. and Farbiszewski, R. (1997a) Antioxidant status of liver, erythrocytes, and blood serum of rats in acute methanol intoxication. *Alcohol (Fayetteville, N.Y.)* **14**, 431–437.

Skrzydlewska, E. and Farbiszewski, R. (1997b) Decreased antioxidant defense mechanisms in rat liver after methanol intoxication. *Free Radic. Res.* **27**, 369–375.

Skrzydlewska, E. and Farbiszewski, R. (1997c) Glutathione consumption and inactivation of glutathione-related enzymes in liver, erythrocytes and serum of rats after methanol intoxication. *Arch. Toxicol.* **71**, 741–745.

Skrzydlewska, E. and Farbiszewski, R. (1997d) Trolox-derivative antioxidant protects against methanol-induced damage. *Fundam. Clin. Pharmacol.* **11**, 460–465.

Skrzydlewska, E., Witek, A., and Farbiszewski, R. (1998) The comparison of the antioxidant defense potential of brain to liver of rats after methanol ingestion. *Comp. Biochem. Physiol. C Pharmacol. Toxicol. Endocrinol.* **120**, 289–294.

Smith, E.N. and Taylor, R.T. (1982) Acute toxicity of methanol in the folate-deficient acatalasemic mouse. *Toxicology* **25**, 271–287.

Soffritti, M., Belpoggi, F., Cevolani, D., Guarino, M., Padovani, M., and Maltoni, C. (2002) Results of long-term experimental studies on the carcinogenicity of methyl alcohol and ethyl alcohol in rats. *Ann. NY. Acad. Sci.* **982**, 46–69.

Soffritti, M., Belpoggi, F., Tibaldi, E., Esposti, D.D., and Lauriola, M. (2007) Life-span exposure to low doses of aspartame beginning during prenatal life increases cancer effects in rats. *Environ. Health Perspect.* **115**, 1293–1297.

Sokoro, A.A., Zhang, Z., Eichhorst, J.C., Zello, G.A., House, J.D., Alcorn, J., and Lehotay, D.C. (2008) Formate pharmacokinetics during formate administration in folate-deficient young swine. *Metabolism* **57**, 920–926.

Spiteller, G. (1998) Linoleic acid peroxidation--the dominant lipid peroxidation process in low density lipoprotein--and its relationship to chronic diseases. *Chem. Phys. Lipids.* **95**, 105–162.

Stanton, M.E., Crofton, K.M., Gray, L.E., Gordon, C.J., Boyes, W.K., Mole, M.L., Peele, D.B., and Bushnell, P.J. (1995) Assessment of offspring development and behavior following gestational exposure to inhaled methanol in the rat. *Fundam. Appl. Toxicol.* **28**, 100–110.

Suit, P.F. and Estes, M.L. (1990) Methanol intoxication: clinical features and differential diagnosis. *Cleve. Clin. J. Med.* **57**, 464–471.

Sweeting, J.N., Siu, M., McCallum, G.P., Miller, L., and Wells, P.G. (2010) Species differences in methanol and formic acid pharmacokinetics in mice, rabbits and primates. *Toxicol. Appl. Pharmacol.* **247**, 28–35.

Sweeting, J.N., Siu, M., Wiley , M.J., and Wells, P.G. (2011) Species- and strain-dependent teratogenicity of methanol in rabbits and mice. *Reprod. Toxicol. (Elmsford, N.Y.)* **31**, 50–58.

Sweeting, J.N. and Wells, P.G. (2010) Species differences in catalase and alcohol dehydrogenase activity in mice and rabbits [ABSTRACT]. *Toxicologist* **114**, 177; PS 834.

Teng, S., Beard, K., Pourahmad, J., Moridani, M., Easson, E., Poon, R., and O'Brien, P.J. (2001) The formaldehyde metabolic detoxification enzyme systems and molecular cytotoxic mechanism in isolated rat hepatocytes. *Chem. Biol. Interact.* **130–132**, 285–296.

Tephly, T.R., Parks, R.E., Jr, and Mannering, G.J. (1964) Methanol metabolism in the rat. *J. Pharmacol. Exp. Ther.* **143**, 292–300.

Teschke, R., Hasumura, Y., and Lieber, C.S. (1974) NADPH-dependent oxidation of methanol, ethanol, propanol and butanol by hepatic microsomes. *Biochem. Biophys. Res. Commun.* **60**, 851–857.

Thannickal, V.J. and Fanburg, B.L. (2000) Reactive oxygen species in cell signaling. *Am. J. Physiol. Lung Cell Mol. Physiol.* **279**, L1005–1028.

Uchida, K. and Stadtman, E.R. (1992) Modification of histidine residues in proteins by reaction with 4-hydroxynonenal. *PNAS* **89**, 4544–4548.

Uotila, L. and Koivusalo, M. (1974) Formaldehyde dehydrogenase from human liver. Purification, properties, and evidence for the formation of glutathione thiol esters by the enzyme. *J. Biol. Chem.* **249**, 7653–7663.

US EPA (2009) Toxicological Review of Methanol in Support of Summary Information on the Integrated Risk Information System (IRIS) (External Peer Review Draft). EPA/635/R-09/013.

Vetrano, A.M., Heck, D.E., Mariano, T.M., Mishin, V., Laskin, D.L., and Laskin, J.D. (2005) Characterization of the oxidase activity in mammalian catalase. *J. Biol. Chem.* **280**, 35372–35381.

Vieira, I., Sonnier, M., and Cresteil, T. (1996) Developmental expression of CYP2E1 in the human liver. Hypermethylation control of gene expression during the neonatal period. *Eur. J. Biochem./FEBS* **238**, 476–483.

Von Burg, R. (1994) Methanol. *J. Appl. Toxicol.* **14**, 309–313.

Wagner, F.W., Burger, A.R., and Vallee, B.L. (1983) Kinetic properties of human liver alcohol dehydrogenase: oxidation of alcohols by class I isoenzymes. *Biochemistry.* **22**, 1857–1863.

Wang, M., Cheng, G., Villalta, P., Kassie, F., and Hecht, S. (2008) Analysis of formaldehyde-DNA adducts in leukocyte DNA by liquid chromatography-electrospray ionization-tandem mass spectrometry: application to rats treated with aspartame. *AARC, San Diego, CA.* **Abstract 2250**.

Wang, Z., Chandrasena, E.R., Yuan, Y., Peng, K.W., van Breemen, R.B., Thatcher, G.R., and Bolton, J.L. (2010) Redox cycling of catechol estrogens generating apurinic/apyrimidinic sites and 8-oxo-deoxyguanosine via reactive oxygen species differentiates equine and human estrogens. *Chem. Res. Toxicol.* **23**, 1365–1373.

Ward, K.W., Perkins, R.A., Kawagoe, J.L., and Pollack, G.M. (1995) Comparative toxicokinetics of methanol in the female mouse and rat. *Fundam. Appl. Toxicol.* **26**, 258–264.

Webster, W.S., Walsh, D.A., McEwen, S.E., and Lipson, A.H. (1983) Some teratogenic properties of ethanol and acetaldehyde in C57BL/6J mice: implications for the study of the fetal alcohol syndrome. *Teratology* **27**, 231–243.

Wells, P.G., Bhuller, Y., Chen, C.S., Jeng, W., Kasapinovic, S., Kennedy, J.C., Kim, P.M., Laposa, R.R., McCallum, G.P., Nicol, C.J., Parman, T., Wiley, M.J., and Wong, A.W. (2005) Molecular and biochemical mechanisms in teratogenesis involving reactive oxygen species. *Toxicol. Appl. Pharmacol.* **207**, 354–366.

Wells, P.G., Kim, P.M., Laposa, R.R., Nicol, C.J., Parman, T., and Winn, L.M. (1997) Oxidative damage in chemical teratogenesis. *Mutat. Res.* **396**, 65–78.

Wells, P.G., Lee, C.J.J., McCallum, G.P., Perstin, J., and Harper, P.A. (Eds.) (2009a) *Chapter 6. Receptor- and reactive intermediate-mediated mechanisms of teratogenesis*, Vol. **196**, Springer, Heidelberg, pp. 131–162.

Wells, P.G., McCallum, G.P., Chen, C.S., Henderson, J.T., Lee, C.J., Perstin, J., Preston, T.J., Wiley, M.J., and Wong, A.W. (2009b) Oxidative stress in developmental origins of disease: teratogenesis, neurodevelopmental deficits, and cancer. *Toxicol. Sci.* **108**, 4–18.

Wells, P.G., McCallum, G.P., Lam, K.C., Henderson, J.T., and Ondovcik, S.L. (2010) Oxidative DNA damage and repair in teratogenesis and neurodevelopmental deficits. *Birth Defects Res. C Embryo Today* **90**, 103–109.

Weston, W.M., Greene, R.M., Uberti, M., and Pisano, M.M. (1994) Ethanol effects on embryonic craniofacial growth and development: implications for study of the fetal alcohol syndrome. *Alcohol. Clin. Exp. Res.* **18**, 177–182.

Winn, L.M. and Wells, P.G. (1999) Maternal administration of superoxide dismutase and catalase in phenytoin teratogenicity. *Free Radic. Biol. Med.* **26**, 266–274.

Wong, A.W., McCallum, G.P., Jeng, W., and Wells, P.G. (2008) Oxoguanine glycosylase 1 protects against methamphetamine-enhanced fetal brain oxidative DNA damage and neurodevelopmental deficits. *J. Neurosci.* **28**, 9047–9054.

Xie, Y., Yang, H., Cunanan, C., Okamoto, K., Shibata, D., Pan, J., Barnes, D.E., Lindahl, T., McIlhatton, M., Fishel, R., and Miller, J.H. (2004) Deficiencies in mouse Myh and Ogg1 result in tumor predisposition and G to T mutations in codon 12 of the K-ras oncogene in lung tumors. *Cancer Res.* **64**, 3096–3102.

Youssef, A.F., Baggs, R.B., Weiss, B., and Miller, R.K. (1997) Teratogenicity of methanol following a single oral dose in Long-Evans rats. *Reprod. Toxicol. (Elmsford, N.Y.)* **11**, 503–510.

8 Methanol and Cancer

JOHN J. CLARY

Bio Risk, Midland, MI,USA

8.1 INTRODUCTION

Methanol has been a commercial chemical for more than 100 years and there is no clear evidence in humans or animals that suggests methanol is a carcinogenic risk. There have been no epidemiological studies in humans and only several limited bioassays in rodents that could provide information about methanol potential carcinogenicity. Rodent studies include exposure by skin painting, inhalation, and by the oral route. None are up to the "gold" standard of a NTP bioassay. Methanol is a high volume chemical that was placed on the NTP list of candidate chemicals for a bioassay years ago in rodents. The testing was not conducted by NTP because of the low interest and the fact that they had judged that rodents were not appropriate model for man in the case of methanol (Masten, 2006). Methanol is not genotoxic, which may have been a factor in the low interest in carrying out a bioassay on methanol.

These rodent studies are presented here in this chapter along with possible mechanism of action of methanol such as genotoxic or oxidative damage that may be factors in the response seen in these studies.

The Toxicology of Methanol, First Edition. Edited by John J. Clary.

8.2 RODENT BIOASSAY

8.2.1 Oral

8.2.1.1 Rats The Ramazzini Foundation conducted a lifetime cancer study in rats (Soffritti et al., 2002). Methanol was administered in the drinking water to Sprague-Dawley rats at 0, 500, 5000, or 20,000 ppm. The study started with the rats at 8 weeks of age and continued until the all rats died. There were 100 Sprague-Dawley rats per sex per group. No testing guideline nor was compliance with GLP reported. The methanol used was 99.8% pure and was replaced every 3 months. Survival, body weight, feed, and water consumption were reported to have been measured, but no data was presented in the published paper. It was concluded that methanol treatment had no effect on survival, body weight, feed, and water consumption (this actual data is presented on the Ramazzini Foundation website). No clinical tests were conducted and blood methanol was not determined. Histopathological evaluation of tissues was conducted when the animals died.

No organ weight, or macroscopic, or non tumor data were reported. Tumors were the only histopathological evaluation reported (significant treatment related effects in the tumor response).

A positive dose–response relationship with a statically significant trend was noted in (1) hemolymphorecticular tumors (all treated females), (2) in tumors bearing animals, and (3) in carcinomas of head and neck (males all levels, females top two doses).

A statistically significant increase in hemolymphorecticular tumors was reported in the high dose females. (Hemolymphorecticular tumors included thymus, spleen, mediastinal lymph nodes, and mesentic lymph nodes and were not found in other tissues.) A statistically significant increase in carcinomas of ear duct in top dose males and tumor-bearing animals—top dose both sexes were also reported (see Table 8.1).

The authors of this study concluded: "Methanol must be considered a multipotential carcinogenic agent."

The actual reported incident rate of hemolymphorecticular tumors in all treated groups plus control rate is seen in Tables 8.2–8.7. Hemolymphorecticular tumors include thymus, spleen, mediastinal lymph nodes,

TABLE 8.1 Key Significant Cancer Findings in Ramazzini Foundation Methanol Study

Methanol, (ppm)	Sex (Number of Animals)	Total Malignant Tumors	Ear Duct Carcinomas	Hemolymphoreticular Neoplasias
20,000	M (100)	70*	24*	40
20,000	F (100)	63*	19	28**
5,000	M (100)	64	17	36
5,000	F (100)	48	16	24
500	M (100)	55	13	35
500	F (100)	48	8	24
0	M (100)	50	9	28
0	F (100)	43	9	13

*$p < 0.01$ using X^2 test.
**$p < 0.05$ using X^2 test.

and mesentic lymph nodes and are not found in other tissues. It is not a common or appropriate practice to combine histiocytic sarcoma, monocyclic leukemia, or myeloid leukemia with neoplasm of lymphocytic origin as done by the Ramazzini Foundation. It should also be pointed out that diagnosis of lesions including lymphoma and leukemia might be difficult if disease or autolysis is a factor. In the males, the concurrent hemolymphorecticular tumor rate is higher than the historical rate but still

TABLE 8.2 Key Significant Cancer Findings in Male Rats in Ramazzini Foundation Methanol Study

Methanol, (ppm)	Sex (Number of Animals)	Total Malignant Tumors	Ear Duct Carcinomas	Hemolymphoreticular Neoplasias
20,000	M (100)	70*	24*	40
5,000	M (100)	64	17	36
500	M (100)	55	13	35
0	M (100)	50	9	28

*$p < 0.01$ using X^2 test.

TABLE 8.3 Key Significant Cancer Findings in Female Rats in Ramazzini Foundation Methanol Study

Methanol, (ppm)	Sex (Number of Animals)	Total Malignant Tumors	Ear Duct Carcinomas	Hemolymphoreticular Neoplasias
20,000	F (100)	63*	19	28**
5,000	F (100)	48	16	24
500	F (100)	48	8	24
0	F (100)	43	9	13

*$p < 0.01$ using X^2 test.
**$p < 0.05$ using X^2 test.

within the normal historical range reported for the laboratory. No positive dose–response relationship with a statically significant trend was reported in males. In the females, the concurrent control hemolymphorecticular tumors rate is the same as the historical control rate. Only the top dose female rate had a statically significant increased of hemolymphorecticular tumors over the control. A positive dose–response relationship with a statically significant trend in hemolymphorecticular tumors was

TABLE 8.4 Dose–Response Trends in Ramazzini Foundation Methanol Study

Methanol, (ppm)	Sex (Number of Animals)	Total Malignant Tumors	Ear Duct Carcinomas	Hemolymphoreticular Neoplasias
20,000	M (100)	70*	24*	40
20,000	F (100)	63*	19*	28**
5,000	M (100)	64*	17*	36
5,000	F (100)	48*	16*	24**
500	M (100)	55*	13*	35
500	F (100)	48**	8*	24**
0	M (100)	50	9	28
0	F (100)	43	9	13

*$p < 0.01$ Cochrane–Armitage test for dose–response relationship.
**$p < 0.05$ Cochrane–Armitage test for dose–response relationship.

TABLE 8.5 Dose–Response Trends in Male Rats in Ramazzini Foundation Methanol Study

Methanol, (ppm)	Sex (Number of Animals)	Total Malignant Tumors	Ear Duct Carcinomas	Hemolymphoreticular Neoplasias
20,000	M (100)	70*	24*	40
5,000	M (100)	64*	17*	36
500	M (100)	55*	13*	35
0	M (100)	50	9	28

$*p < 0.01$ Cochrane–Armitage test for dose–response relationship.

reported in females. The hemolymphorecticular tumor rate in the historical controls ranges to as high as 25%, so only the top dose is slightly higher than the top end of historical controls. The top dose is 40 times higher than the low dose and the increase in hemolymphorecticular tumors rate in the top dose is only 4% higher than the low dose. Not much of an increase in hemolymphorecticular tumors rate given the wide dose range employed in this study.

This study is not the typical design and some questions require further discussion. It has been suggested that by allowing animal to live out their life, as done in this study, treatment-related tumors that only appear in old age can be detected that would be missed in a standard

TABLE 8.6 Dose–Response Trends in Female Rats in Ramazzini Foundation Methanol Study

Methanol, (ppm)	Sex (Number of Animals)	Total Malignant Tumors	Ear Duct Carcinomas	Hemolymphoreticular Neoplasias
20,000	F (100)	63*	19*	28**
5,000	F (100)	48*	16*	24**
500	F (100)	48**	8*	24**
0	F (100)	43	9	13

$*p < 0.01$ Cochrane–Armitage test for dose–response relationship.
$**p < 0.05$ Cochrane–Armitage test for dose–response relationship.

TABLE 8.7 Hemolymphoreticular Neoplasias by Type

Methanol, ppm	Sex (Number of Animals)	Lymphoblastic Lymphoma	Lymphoctic Lymphoma	Lymphoimmunoblastic Lymphoma	Histocytic Sacroma Monocytic Leukemia	Myeloid Leukemia	Total
20,000	M (100)	1	0	37	1	1	40
20,000	F (100)	1	0	21	3	3	28*' **
5,000	M (100)	1	0	28	1	6	36
5,000	F (100)	1	0	19	2	3	24**
500	M (100)	3	0	24	4	4	35
500	F (100)	1	1	17	2	3	24**
0	M (100)	1	0	16	3	8	28
0	F (100)	0	0	9	1	3	13

Historical for laboratory.

*$p < 0.05$ using X^2 test.

**$p < 0.05$ Cochrane–Armitage test for dose–response relationship.

2-year bioassay. There are not much published data on historical tumor rates in rats over 2 years of age. Ramazzini Foundation has 20 years of data on about 2000 control males and females that were allowed to live until spontaneous death. This would be useful to gets time to historical tumor data especially after 2 years. An increase in leukemia/lymphoma was reported by Ramazzini Foundation using the lifetime study design in at least seven chemicals, but not in the other studies in the literature of the same chemicals, which used a standard 2-year design (Clary, 2007). In one of the chemicals (MTBE), the top dose males lived significantly longer than the control males but had a lower incidence of leukemia/lymphoma than the control (Belopoggi et al., 1997). Background tumor data in animals more than 2 years old is limited. Increase in normal old age tumors can also be a confounding factor.

In a review of the Ramazzini Foundation aspartame study by EFSA (2006), chronic respiratory disease and poor survival was reported; in their evaluation survival ranged from 22% to 45% at 103 weeks. EFSA concluded that the increased incidence of lymphomas/leukemias reported in treated rats was unrelated to aspartame, given the high background incidence of chronic inflammatory changes in the lungs and the lack of a positive dose–response relationship. It is possible that questionable diagnosis of some tumors can arise as a result of abundant lymphoid hyperplasia in the lungs of rats suffering from chronic respiratory disease.

A review of the possible role of *Mycoplasma pulmonis* and diagnosis of lymphomas in the Ramazzini Foundation aspartame, MTBE and methanol studies were conducted (Schoeb et al., 2009). In all three studies, the most commonly reported hematopoietic tumor was lymphoimmunoblastic lymphomas. It was found in high percentages in the lung and as many as 50% of the time lymphoimmunoblastic lymphomas were reported only in the lung and not at other site. In the Ramazzini Foundation methanol study, the incidence of lymphoimmunoblastic lymphomas reported only in the lung ranging from 33.3% to 56.3% for all groups Lymphoimmunoblastic lymphomas are normal a rare tumor in Sprague Dawley rats. In the lung, neurtrophils, plasma cell, and lymphocytes accumulation are typically seen following an infection of *M. pulmonis* rather than a rare form of lymphomas. Schoeb et al. (2009)

concluded that the lesions in the lung are most likely caused by *M. pulmonis*, which may have been incorrectly interpreted as lymphomas by the Ramazzini Foundation (Schoeb et al., 2009).

Correct diagnosis is also a question. In the Ramazzini Foundation study of aspartame, 75 slides from the study were reviewed by a National Toxicity Program pathology working group (PWG) (NTP, 2004). On the basis of this review, it appears that the Ramazzini Foundation "appeared to apply a more severe classification to lesions than the NTP PWG." For example, the NTP PWG confirmed the diagnoses of only two out of six ear duct carcinoma reported by Ramazzini Foundation in this limited review. Over classification by Ramazzini Foundation of tumors in the mammary gland, pelvic and oral cavities were also suggested in the aspartame study by the NTP (2004). The Ramazzini Foundation also reported a treatment related increase in total tumors. An over classification by Ramazzini Foundation would also affect the total number of tumor-bearing animals.

In a later partial review of the Ramazzini Foundation, methanol study by a team of NTP pathologists was not sufficient to support or refute the conclusions of Ramazzini Foundation (NTP, 2010). The NTP pathologists reviewed histological sections of the lung, spleen, lymph nodes, thymus, ear canal, Zymbal's gland, and cranium from the 100 high dose and 100 control males in the Ramazzini Foundation methanol study. In the lung, the NTP pathologist reported a higher incidence of lung inflammation and a lower frequency of leukemia or lymphomas than was reported by the Ramazzini Foundation. NTP pathologists reported inflammation in the nose of most animals as well as inflammation in the inner ear, but reported half the incidence of squamous cell carcinomas in the ear that were reported by the Ramazzini Foundation. The NTP pathologists also reported fewer osteosarcomas in the cranium than did the Ramazzini Foundation in this comparison. They also reported fewer leukemia or lymphomas overall than did the Ramazzini Foundation. The NTP pathologists noted no significant differences, when the results were compared to the Ramazzini Foundation report, in liver, spleen, thymus, or Zymbal's gland tumors.

In 2011, a detailed pathology review (pathology data review, quality assurance, and PWG) of five Ramazzini studies was conducted at the request of the NTP and EPA (NTP & EPA, 2011). Methanol was one of the five studies reviewed. The PWG reviewed 744 slides from 367 rats in the methanol study. These slides were from the rats diagnosis by the original study pathologists to have lymphoma or leukemia, as well as, animals reported to have ear or bone neoplasm. The PWG reported fewer lymphoid neoplasms primarily from the respiratory tract as well as fewer neoplasms in the inner ear and cranium. Chronic inflammation was noted in the respiratory tract and ear suggesting that respiratory infections were present. Chronic infection was thought by the PWG to cause hyperplasia that was misdiagnosed in many cases by the original study pathologists as malignant neoplasms. In areas where chronic infection was not seen, the PWG generally agreed with original study pathologists diagnosis of tumor types, location, and numbers. The PWG concluded that there were fewer malignant neoplasms than reported by the original study pathologists. As a result of this review, EPA has announced that it will not use cancer data from this study in its ongoing risk assessment of methanol because of the overestimating of cancer incidence by the original study pathologists (EPA, 2012).

Autolytic changes can also be a problem in interpreting lesions when animals are not scarified in extremis or at a specific time in the study. Autolytic changes were also a confounding finding of the NTP (2004) in the Ramazzini Foundation aspartame study and an occasionally precluded histological diagnosis by the NTP pathologists reviewing the methanol study (NTP, 2010). In neither the Ramazzini Foundation aspartame study nor methanol study was all the study data available for review by the working group. Ward and Alden (2009) questioned the qualification and accuracies of the pathologist diagnosis in the Ramazzini Foundation aspartame study and pointed out that without all the data being available, a rigorous review for regulatory purposes of the study is controversial.

8.2.1.2 Mice A study to evaluate the toxicity and carcinogenicity of malonaldehyde was conducted at the Eppley Institute (Apaja, 1980). Malonaldehyde (MDA) can only be kept for a short period of time, as it is unstable and hydroscopic. MDA was prepared for this study as a dilute solution which was made more stable than pure MDA, by dissolving malondialdehyde bis diethylacetal and 6 N HCL in water. Methanol was released by the hydrolysis of the acetal groups and MDA was produced in the drinking water. Mice received 0.5%, 0.25%, or 0.125% (MDA) in drinking water 6 days a week for life. Three methanol control groups were also included in the study. Methanol ("controls") was added to water to create control solutions at levels that represented the stoichiometric amount of methanol liberated by the hydrolysis of the acetal group in the different MA test groups. Six groups of 25 females and six groups 25 males were used in this study (three treatment groups and three methanol groups per sex). The mice were observed daily, weighed weekly, and sacrificed when moribund. Complete autopsy were conducted and section of skin, lung, liver, spleen, pancreas, kidney, adrenal glands, esophagus, stomach, small and large intestine, rectum urinary bladder, uterus, ovaries, or testes, prostate gland vesicular glands, and any gross lesions, including tumors, were evaluated.

There was a treatment-related decrease in water consumption in the MDA groups when compared to methanol control. The calculated methanol daily dose in the MDA-treated groups ranged from 49.8 mg in the top dose males to 15.3 mg in the low dose females. The top male treatment group actually received on average 28 mg MDA per day, while the low dose female received on average 8.6 mg MDA per day based on the reduced treated water intake. On the basis of a calculated daily methanol dose in the MDA-treated group, the methanol dose ranged from 450 to 2000 mg/kg. Saturation of catalase, the enzyme that biotransforms methanol to formaldehyde is reported to start at about 600 mg-methanol/kg bw in rats (Horton et al., 1992). Oxidative damage is reported when catalase is saturated in rats (Skrzydlewska and Farbiszewski, 1996, 1998; Skrzydlewska et al., 1998; Parthasarathy et al., 2006; Dobrzynska et al., 1999). One of the products of this oxidative damage is MDA. So it is possible that the MDA dose in the mid and top MDA groups was actually higher than the calculated MDA levels because of additional production of MDA as a result of lipid peroxidation induced by methanol.

In the methanol control groups, the dose of methanol was much higher than the actual dose in the respective MDA-treated animals because of decreased water consumption in the MDA groups. The calculated level of methanol ingested was 84.5/82.7 mg in the high dose female/male methanol only group, respectively. The level of methanol in the low methanol only group was 22.6/24.6 mg (female/male, respectively). Survival of both male and female in the MDA top dose was significantly lower than the corresponding methanol control groups. There was a difference in body weight between the top two MDA dose groups and the corresponding methanol controls. There were no differences in body weight between low dose MDA groups and corresponding methanol control.

The top female MDA group and the mid dose males had lower lung tumors incidence than the corresponding methanol control. The incidence of blood vessel tumor was not affected by treatment. In the top two dose methanol control groups there was a significantly larger number of malignant lymphomas than in the corresponding MA groups. Table 8.8 gives a summary of malignant lymphomas in female mice and methanol dose in each of the six groups. The lymphoma historical control rate was 18% at Eppley, but was reported to vary from study to study. It should be noted that no water only control group was included in this experiment. It is interesting to note that in the female mice receiving just methanol, a dose-related increase in malignant lymphomas was noted (Table 8.8), but no trend was seen in the male mice receiving methanol only (Table 8.9). Also of interest is that an inverse dose relationship increase in malignant lymphomas was noted in both

TABLE 8.8 Lymphomas Versus Methanol Concentration (Female Mice)

Treatment Group	Average Daily Methanol Dose, mg[a]	Average Methanol, mg/kg bw[b]	Percentage Lymphomas
0.899% methanol	84.5	2,400	40
0.444% methanol	40.8	1,160	36
0.222% methanol	22.6	645	16
0.5% MDA	41.8	1,190	4
0.25% MDA	26.2	750	28
0.125% MDA	15.3	440	36
Historical control	0	0	18

[a]Average daily methanol dose based on drinking water consumption.
[b]Calculated based on an average weight of 35 g.

TABLE 8.9 Lymphomas Versus Methanol Concentration (Male Mice)

Treatment Group	Average Daily Methanol Dose, mg[a]	Average, mg/kg bw[b]	Percentage Lymphomas
0.899% Methanol	82.7	2,050	16.7
0.444% Methanol	43.5	1,090	24
0.222% Methanol	24.6	615	4
0.5% MDA	49.8	1,245	0
0.25% MDA	31.5	790	12.5
0.125% MDA	18.2	455	25
Historical control			

[a]Average daily methanol dose based on drinking water consumption.
[b]Calculated based on an average weight of 40 g.

the males and females receiving MDA (Tables 8.8 and 8.9). The top dose females receiving MDA had a 4% rate of malignant lymphomas and the lymphomas rate in the top dose MDA male rate was 0. The variable malignant lymphomas rate from group to group was judged to be a random observation by the author (Apaja, 1980).

In a short-term study in mice treated with 2 g of methanol/kg i.p., no increase in 8-oxo-7, 8-dihydro-2′-deoxyguanosine (8-oxodG), a frequent marker of oxidative DNA damage was seen, but a modest increase in HNE–His protein adducts, a product of increased lipid peroxidation was noted in the mouse bone marrow (McCallum et al., 2011a, b). This suggests that the dose increase in lymphomas in female mice in this study by an oxidative DNA damage is unlikely, but the question about any role of HNE–His protein adducts in causing lymphomas is open. It is not likely that methanol would cause lymphomas or other cancers in humans as a result of oxidative damage, because oxidative damage is an unlikely mechanism of action in humans exposed to methanol (see chapter 4).

8.2.2 Inhalation

8.2.2.1 Rats In 1986 and 1987, the New Energy Development Organization reported on a very large project entitled *Toxicological Research* *of Methanol as a Fuel for Power Station* (NEDO, 1986, 1987). The NEDO studies were a very large undertaking, conducted to evaluate methanol as a fuel for power stations. Theses reports, which were not published, reviewed studies of methanol conducted in monkey, rats, mice, and aquatic species. These inhalation studies included acute, chronic, carcinogenicity and effects on reproduction and the fetuses. The first phase of the research was made available in English in a 114-page summary (NEDO, 1986). The research in monkeys, rats, and mice were made available in the second summary of 300 pages (NEDO, 1987). There is a great deal of information provided in the summary, in some cases as much as you would find in many published papers. In these studies, there was a 24-month inhalation study in rats and a 18-month inhalation study in mice to assess carcinogenicity of methanol. Unfortunately, the results of these two inhalation studies were never published by NEDO in a peer-reviewed journal. A review of these two old inhalation studies has been recently published (Cruzan, 2009).

In the rat carcinogenic study, exposure was by inhalation for 20 hours per day, 7 days per week for 2 years with concentrations of methanol of 0, 10, 100, or 1000 ppm. Fischer 344 rats were used, 52/sex per group. OECD testing guidelines were used, but no indication of GLP was followed. NEDO collected clinical tests data (urine, hematological and biochemical), body weight, feed consumption, and organ weights in addition to macroscopic and histological evaluations of the standard tissues. Blood methanol was also measured.

No treatment-related effects were observed in clinical tests, body weight, feed consumption, and organ weights. The lack of effect on body weight indicates that the top dose may have been below a maximum tolerated dose (MTD). Survival was 60–80% at termination (104 weeks). Blood methanol was elevated in the high dose males and females (M 54 ppm, F 88 ppm), but not at the other lower doses. The elevation of blood methanol indicates the beginning of saturation of catalase (600–1000 mg/kg), the key enzyme that metabolizes methanol to formaldehyde in rodents. On the basis of the blood levels reported the top dose of methanol was close to 600 mg/kg body weight.

There was an increase in nodes in the lung of the high dose males with most of the nodes being papillary adenomas. In male rats lung nodes were seen in 2/52, 4/52, 5/52, and 10/52 at concentrations of 0, 10, 100, and 1000 ppm methanol, respectively. At the top dose methanol concentration (1000 ppm) the increase in nodes is statically significant.

Lung papillary adenomas in males were seen in 1/52, 5/50, 2/52, and 6/52 at concentrations of 0, 10, 100, and 1000 ppm methanol, respectively. The increase in lung papillary adenomas was not significantly different from the historical control rate.

The authors of the study concluded that " methanol did not induce any carcinogenic or toxic effects in rats at dose level of 1000 ppm or less."

8.2.2.2 Mice The design of the mouse inhalation study conducted by NEDO (1986, 1987) was very similar in design to the inhalation rat study just discussed. The major difference was the study ran for 18 months instead of 2 years as in the rat study. Exposure was by inhalation for 19 hours per day, 7 days per week for 18 months with concentrations of methanol of 0, 10, 100, or 1000 ppm. CRJ:B6C3F mice were used, 52 males and 53 females per group. OECD testing guidelines were used, and again no indication that the study was conducted under GLP. Clinical tests, including checking for infection, body weight, feed consumption, and organ weights were measured in addition to macroscopic and histological evaluations of the standard tissues were evaluated. Survival was 93–100% at termination (78 weeks). Blood methanol was not measured.

No treatment-related effects were observed in clinical tests, body weight, feed consumption, organ weights, or histopathological examinations. The lack of effect on body weight indicates that the top dose may have been below a maximum tolerated dose (MTD). The authors of the study concluded that "methanol did not induce any carcinogenic or toxic effects in mice at dose level of 1000 ppm or less."

8.2.3 Dermal

8.2.3.1 Mice Methanol was used as a control/solvent in a topical application study of photocarcinogenesis of retinoic acid. Retinoic acid was dissolved in methanol for application to the back of 30 males and 30 females hairless mice (20 cm^2) daily for 30 weeks. A control group received 100 ul methanol only daily for 30 weeks. In a second group of

30 males and 30 females hairless mice, the mice received UVR (2 hours of simulated sunlight) in addition to methanol. The animals were observed until 55 weeks of age. No skin tumors were observed in the methanol only treated group at 55 weeks. Twenty-three animals treated with methanol and UVR had skin tumors (carcinomas) out of 51 survivors at 55 weeks. Mice receiving retinoic acid and UVR or croton oil plus retinoic acid had closed to 100% of the survivors with skin tumors (multiple tumors per animal). Limited necropsies were preformed. Hyperplasic, but no metastases, in the lymph nodes, spleen, and salivary gland were noted in animals with tumors. The skin of the methanol only treated mice and the methanol plus UVR-treated mice appeared normal. Methanol by itself was not a tumor producer, but methanol with 2 hours of simulated sunlight produced a tumor rate of 45%. Retinoic acid treatment on the other hand, produced multiple tumors per animal in about 100% of the mice (Forbes et al., 1979).

A study to evaluate the dermal toxicity and carcinogenicity of malonaldehyde was conducted at the Eppley Institute (Apaja, 1980). Malonaldehyde (MDA), which is a product of oxidative lipid deterioration in rancid beef, had previously reported to be carcinogenic in a two-stage skin-painting study using acetone as a solvent. Malonaldehyde was found to be too unstable in acetone, so methanol, which helped stabilize MDA, was used in the Eppley study. MDA was prepared for this study as a dilute solution, that is more stable than pure MDA, by dissolving malonaldehyde bis diethylacetal and 6 N HCL in water. Methanol was released by the hydrolysis of the acetal groups and MDA was produced. In the Eppley skin-painting study, the control group was treated with 0.05 ml of methanol (containing 0.625% methanol in water). Groups of 40 female shaved Swiss mice were treated with various levels of MDA and methanol control three times a week for life. The mice were observed daily (dermal irritation), weighed weekly, and sacrificed when moribund. Complete autopsy were conducted and section of skin, lung, liver, spleen, pancreas, kidney, adrenal glands, esophagus, stomach, small and large intestine, rectum urinary bladder, uterus and ovaries, vesicular glands, and any gross lesions including tumors.

There was no treatment-related skin irritation in any group. Because of lower than expected doses in the lower MDA groups due to technical problems, only the top MDA group (1.2%) and the methanol control were kept for a lifetime. There was no difference in survival or body weight between the top dose MA group and the methanol control. No skin tumors were observed, but the top MDA group had a significantly larger number of malignant lymphomas (17) than the methanol controls (6). The lymphoma incidence was within the normal historical range at Eppley. It should be noted that no water only control group was included in this experiment. There were no differences between the top dose MDA group and the methanol control in any other tumor incidences (Apaja, 1980).

8.3 POSSIBLE MECHANISMS

8.3.1 Genotoxicity

8.3.1.1 In Vitro Methanol did not show activity in the Salmonella test with or without metabolic activation in Salmonella strain TA 1535, TA 1537, TA 1538, TA 98, and TA 100 in several studies (Simmon et al., 1977; Deflora, 1981; NEDO, 1986). Methanol is also negative in *Escherichia coli* (WP2 WP 67 CM 871), Spombe ade 6 locus and *Neurospora crassa* tests with and without activation (IPCS, 1997). Methanol did not show activity in *E. coli* (WP2uvrA) (Shimizu et al., 1995). Methanol induced chromosomal malsegregation in *Aspergillus niduians* (Crebelli et al., 1989). Methanol does not cause sister chromatid exchange in Chinese hamster cells (IPCS, 1997; NEDO, 1986). Methanol was also negative in cell transformation assays (Syrian hamster embryo clonal system and Rausher leukemia virus infected rat embryo cells) (IPCS, 1997). Methanol was negative in the L5178Y mouse lymphoma cells without activation, but positive with activation (McGregor et al., 1988). Heidelberger et al. (1983) found methanol negative in Syrian hamster cell (SHE) and did not increase transformation in SHE with and without transformation by simian adenovirus.

8.3.1.2 In Vivo Cytogenic effects of methanol in blood erythrocytes, lung cells, and testicular germ cells of mice were evaluated following exposure to 800 or 4000 ppm for 6 hours per day for 5 days. No increase in frequencies of micronuclei in blood cells, sister-chromatid exchange, chromosome aberrations of micronuclei in lung cells or of synaptonemal complex damage in spermatocytes was noted (Campbell et al., 1991). Methanol did not induce chromosomal aberration, SCE, or increase in micronucleus in tests conducted by NEDO (1986). Methanol was also tested and gave negative results in two human cell line [MCF-7, HepG2] (Ohno et al., 2005). Urine from mice given 5000 mg/kg bw for 5 days showed no mutagenic activity (Chang et al., 1983).

Micronucleus assay was performed in normal and folate-deficient mice, 10 per group, following four daily doses of 0, 300, 600, 1200, or 2500 mg/kg methanol. No indication of genotoxic potential was observed in normal and folate-deficient mice (O'Loughlin et al., 1992).

Methanol was positive in mouse in both chromosomal aberration and micronuclei in erythrocytes assay and in structural chromosome aberration in bone marrow (IPCS, 1997).

Methanol does not appear to be genotoxic based on the majority of *in vitro* and *in vivo* testing reported in the literature. Methanol treatment does not appear to affect blood formaldehyde levels or the rate of breakdown of formaldehyde *in vivo* in animals.

Formaldehyde, the metabolite of methanol is considered a genotoxic *in vitro* but is generally not genotoxic *in vivo* assays.

In a recent elegant study, Lu et al. (2012) has shown that in rats exposed orally for 5 days to 500 mg/kg or 2000 mg/kg of labeled methanol ($^{13}CD_4$) had a dose-related increase in exogenous labeled formaldehyde hydroxymethyl DNA adducts, derived from labeled methanol ($^{13}CD_4$), in multiple tissues (liver, lung, kidney, spleen, thymus, bone marrow, and WBC). The level of exogenous hydroxymethyl formaldehyde DNA adducts derived from methanol was similar to the endogenous formaldehyde hydroxymethyl DNA adducts, normally found in all tissues in the rats dosed at 500 mg/kg. However, in the rats dosed at 2000 mg/kg the level of exogenous formaldehyde hydroxymethyl DNA adducts derived from methanol were higher (~two- to fivefold) than levels found in rats dosed at 500 mg/kg. In the bone marrow, the exogenous formaldehyde hydroxymethyl formaldehyde DNA adduct level was higher than the endogenous formaldehyde hydroxymethyl DNA adducts levels. This study suggests that methanol can form exogenous formaldehyde hydroxymethyl DNA adducts in various tissues in rats that suggests a possible genotoxic mechanism of action. The relevance of this finding to human is unknown because of the metabolic differences between rats and humans.

8.3.2 Oxidative Damage

The mechanism by which methanol causes toxicity in the developing embryo and fetus or possible cancer in rodents is not clearly understood, but a primary or contributing factor may be enhanced "oxidative stress," or increased formation of highly reactive and potentially toxic forms of oxygen termed "reactive oxygen species" (ROS). Oxidative stress and damage occurs when the concentration of ROS generated within a cell or tissue exceeds the antioxidant capacity of the cell or tissues. These ROS include highly toxic "free radical" intermediates such as hydroxyl radical that have been implicated in a number of human diseases and drug toxicities including nongenotoxic cancer, neurodegenerative diseases, atherosclerosis, and birth defects (Roberts et al., 2009). In addition, oxidative damage has been suggested as playing a role in breast, liver, prostate, leukemia, and lung cancer. Tobacco smoking in man, another carcinogenic source of ROS, increases the oxidative DNA damage rate by 35–50% (Loft and Poulsen, 1996). Oxidative stress can be an important part of both initiation and promotion in the cancer process (dieldrin, acrylonitrile) (Jiang et al., 1998; Klaunig et al., 1998; Klaunig and Kamendulis (2004)).

In living cells, ROS are formed continuously as a result of normal metabolism, biochemical reactions, and external factors (chemical exposure). Antioxidant defense systems may not be able to provide complete protection from the effects of increased ROS. These effects include oxidative DNA damage, which has been suggested as an important factor in some carcinogenic responses. The most abundant of these lesions, 8-oxo-7, 8-dihydro-2′-deoxyguanosine (8-oxodG), is the most frequent marker of oxidative DNA damage. Human studies support the notion of oxidative DNA damage as an important mutagenic and apparently carcinogenic factor. In addition, oxidative damage may be important for the aging process, and the pathogenesis of inflammatory diseases (Roberts et al., 2009).

In rats, the oxidation of methanol is provided mainly by catalase (not alcohol dehydrogenase, the primary enzyme in oxidation of alcohols in humans). In humans, catalase is not involved in the metabolism of methanol to formaldehyde and ROS and lipid peroxidation following methanol exposure have not been reported in humans.

Catalase, which is found widely in the rodent tissues, normally provides protection by attacking and detoxifying superoxide radicals such as hydrogen peroxide (Parthasarathy et al., 2006). High doses of methanol saturated the catalase enzyme in rodents (600–1000 mg/kg bw) (Horton et al., 1992) and the blood methanol level increases dramatically as does ROS and lipid peroxidation, which are indicator of oxidative damage. Because methanol is water soluble, it is found throughout the rodent body at doses above saturation of the enzyme catalase. This results in the possibility of lipid peroxidation and ROS in various tissues and organs.

Methanol and oxidative stress has been studied in several short-term studies in rodents, rabbits, and primate. These rat studies are described in greater detail in this book in Chapter 4, while recent studies of methanol and oxidative stress in mice, rabbit, and primates are described in greater detail in Chapter 7. Oxidative stress is seen when the antioxidant capability of the target cell becomes insufficient to prevent the increased formations of ROS. The antioxidant status was evaluated by the increase in lipid peroxidation (thiobarbituric acid reacting substances (TBA)), MDA, and enzymatic changes (superoxide dismutase (SOD), Glutathione peroxidase (GSH-Px), glutathione reductase (GSsG-R), catalase (CAT), and effects on nonenzymatic antioxidants (ascorbate, α-topcopherol, nonprotein- and protein-bound sulfhydryl compounds).

There are several oral short-term studies in rats receiving high doses of methanol (1.5, 3, or 6 g/kg bw which is greater than the level where catalase is saturated) receiving methanol by gavage and were scarified at different time periods from 6 hours to 7 days after dosing.

Following a single oral doses of 1.5 or 3 g/kg bw of methanol was reported to impair the antioxidant defenses in the liver, erythrocyte, and serum (Skrzydlewska and Farbiszewski, 1996, 1998). A decrease in antioxidant activity in the brain and liver following very high oral methanol dosing (6 mg/kg bw) in rats has also been reported (Skrzydlewska et al., 1998).

The effect of methanol on antioxidant status and lipid peroxidation in the lymphoid organs (lymph glands, spleen, thymus, and bone marrow) was investigated following a dose of 2.37 g methanol/kg bw intraperitoneally (i.p.) for 1, 15, or 30 days. Lipid peroxidation was elevated at each time period (higher with increase time of treatment). All enzymatic and nonenzymatic antioxidant indices were significantly elevated following 1-day treatment, but these indices were significantly lower than the control on day 15 and lower still on day 30 indicating oxidative damage to the thymus, spleen, lymph nodes, and bone marrow (Parthasarathy et al., 2006).

Intraperitoneal injects of antioxidants (triolox derivative U-83836E and *N*-acretylcystiene) in rats receiving 3.0 g/kg of methanol orally, partially prevent lipid peroxidation in the erythrocytes demonstrating the protective role of the added antioxidant on oxidative damage caused by methanol (Dobrzynska et al., 1999).

In another study, an increase in the formation of free radicals in Sprague Dawley rats was detected by electron spin resonance spectroscopy following a single dose of methanol (i.p. 4.5 g/kg or gavage 7 g/kg). The same free radical adduct was detected in the bile and urine 2 hours after dosing (Kadiiska and Mason, 2000).

In still another study, hydroxymethyl radicals were detected in rat liver microsomes and nuclei of NADPH-dependent process following treatment with methanol. The results suggest that both rat liver nuclei and microsomes are able to generate free radicals during NADPH-mediated methanol metabolism (Castro et al., 2002).

A study of mice, rabbits, and primates given a single or 15 daily doses of 2.0 g of methanol i.p., (high enough to cause lipid peroxidation in rats) was conducted to assess tissue oxidative DNA damage only. Oxidative DNA damage is most commonly indicated by an increase in 8-oxo-7, 8-dihydro-2′-deoxyguanosine (8-oxodG). 8-oxodG is mutagenic and a potential cause of cancer. No increase in DNA damage (8-oxodG) was noted in the spleen, liver, lung, bone marrow, or kidney in mice, rabbits, or primates given a single dose of 2.0 g of methanol i.p. However, an increase in hydroxynonenal–histidine (HNE–His) protein adducts, an additional measure of lipid peroxidation and free radical damage, were detected in mice bone marrow and rabbit spleen but not in primates in the same study (McCallum et al., 2011a, b). HNE, an end product of lipid peroxidation like MDA may cause mutagenic adducts, and possible effects on DNA (Halliwell and Gutteridge, 2007). Both MDA and HNE have been reported to cause cancer. MDA was found to be carcinogenic to rats but not to mice in a NTP bioassay (NTP, 1988) and HNE levels have been reported to increase in human breast cancer (Kanhtala et al., 2011).

Unfortunately, no data exist on 8-oxodG or HNE–His adduct levels in rats treated with methanol. There are several studies in rats, which indicate that methanol can cause oxidative damage (TBA, MDA, and enzymatic changes and effects on nonenzymatic antioxidants) in rats following high methanol doses (Dobrzynska et al., 1999; Skrzydlewska et al., 1998; Skrzydlewska and Farbiszewski, 1998, 1996). In mice, rabbits, and primates, 8-oxodG or HNE–His adduct data is available but no data on oxidative damage using indicators such as TBA, MDA, enzymatic changes, and effects on nonenzymatic antioxidants end-points are available. Direct comparison of rats and mice response to oxidative damage as a result of methanol exposure is possible. On the basis of similarity in metabolism, it may be assumed that the rat and mouse are similar in their oxidative response to methanol. If this is the case, one cannot rule out that oxidative damage may have played some role in the rats carcinogenic study conducted by the Ramazzini Foundation. In the Ramazzini Foundation lifetime oral methanol drinking water study, the estimated top dose was between 1 and 2 g/kg (Soffritti et al., 2002). In the Eppley drinking water study, methanol was used as a control, the estimated top dose of methanol was >2000 mg/kg (Apaja, 1980). In the NEDO inhalation study, the estimated top dose was about 600 mg/kg based on blood methanol levels (NEDO, 1986, 1987).

Developmental effects are reported in rats and mice at a similar dose possibly as a result of oxidative damage. Wells et al. (1997, 2005, 2009a, b) has suggested that ROS can effect embryonic development in two ways. One effect maybe by altering signal transduction in the embryonic cell; and the second possible way maybe by damage macro-molecules, such as genes and proteins. In the developmental mouse study, the LOEL was 2000 ppm (~1600 mg/kg) (Rogers et al., 1993) and in the developmental rats study the LOAEL was 10,000 ppm (~3700 mg/kg) (Nelson et al., 1985). In these developmental studies, oxidative damage would have been present at the LOEL/LOAEL doses and higher.

It has been shown that methanol causes oxidative stress in rats. Methanol, which is water soluble, is widely distributed throughout the body. High doses of methanol saturated the catalase enzyme in rodents (600–1000 mg/kg); the blood methanol level increases dramatically and there is a likely increase in ROS and oxidative damage throughout the various organ and tissues in the body.

While oxidative DNA damage, based on an increase in 8-oxodG, does not result in mice, rabbits, and primates given a high dose of methanol, one could draw the conclusion that methanol-induced DNA damage would not be observed in rats given a similar dose. MDA, a product of lipid peroxidation, showed clear evidence of carcinogenicity in rats (NTP, 1988). HNE, another product of lipid peroxidation was increased in bone marrow of mice given methanol. One therefore could conclude that products of lipid peroxidation may be a potential mechanism for cancer in rodents.

8.4 HUMAN CANCER DATA

There are no epidemiological studies of cancer in humans exposed primarily to methanol. In other epidemiological studies where methanol may be present with other chemicals, no evidence suggests that methanol is a human cancer risk.

8.5 CONCLUSION

Is methanol a carcinogen in animals or humans? There have been no "gold standard" bioassays of methanol in animals. Methanol is not considered mutagenic and the limited animals studies do not clearly support that methanol is a cancer producing agent in rodents. The doses used in animal studies go beyond the lethal dose in humans and the saturation of catalase, the enzyme that metabolizes methanol to formaldehyde in animals. Blood methanol increases dramatically once catalase is saturated. Methanol is water-soluble so increased methanol is found throughout the tissues in the body. This is not the case in humans, where ADH is the primary enzyme in the breakdown of methanol to formaldehyde. The recent findings (Lu et al., 2012) demonstrate the formation of a dose-related increase in exogenous labeled formaldehyde hydroxymethyl DNA adducts derived from labeled methanol ($^{13}CD_4$) in rats in multiple tissues at and above the lethal dose in humans and the dose above the saturation of catalase suggesting that formaldehyde hydroxymethyl DNA adducts could be a genotoxic factor in the response in rodent bioassays at high doses.

There is also limited evidence that oxidative damage, at high levels above the saturation of catalase, may be a contributor to the some of the responses reported in rat studies. In addition to oxidative damage, the increase in DNA adducts from high doses of methanol and other factors, such as infection and misdiagnosis may be responsible for the reported increase in tumors in some rodent studies. The findings reported in bioassays, if real, appear to have little relevance to humans. The fact that the top doses of methanol used in animal studies, where oxidative damage and DNA adducts formation from methanol occurs, are seen at higher than the acute lethal dose in humans and above the saturation of catalase supports that methanol is unlikely a cancer risk to humans.

REFERENCES

Apaja, M. (1980) Evaluation of toxicity and carcinogenicity of malonaldehyde. Acta Univ. Oul. D 55 *Anat. Path. Microbiol.* **8**, 1–59.

Belopoggi, F., Moranndo, M., Filippini, F., and Maltoni, C. (1997) Results of the long term experimental studies on the carcinogenicity of methyl tert-butyl ether. *Ann. N.Y. Acad. Sci.* **837**, 77–95.

Campbell, J.A., Howard, D.R., Backer, L.C., and Allen, J.W. (1991) Evidence that methanol inhalation does not induce chromosome damage in mice. *Mutat. Res.* **260**, 257–264.

Castro, G.D., Costantini, M.H., Delgado de Layno, A.M., and Castro, J.A. (2002) Rat liver microsomal and nuclear activation of methanol to hydroxymethyl free radicals. *Toxicol. Lett.* **129**, 227–236.

Chang, L.W., McMillian, L., Wynne, B.R., Pereira, M.A., Colley, R.A., Ward, J.B., and Legator, M.S. (1983) The evaluation of six different monitors for exposure to formaldehyde in laboratory animals. *Environ. Mutagen.* **5**, 381.

Clary, J. (2007) Critical review of bioassays on methanol. Proceeding Toxicology Forum Washington DC.

Crebelli, R., Conti, G., Conti, L., and Carere, A. (1989) A comparison study of ethanol and acetaldehyde as a inducers of chromosome malsegregation on *Aspergillus nidulans*. *Mutat. Res.* **215**, 187–195.

Cruzan, G. (2009) Assessment of the cancer potential of methanol. *Crit. Rev. Toxicol.* **39**, 347–363.

Deflora, S. (1981). Study of 106 organic and inorganic compounds in salmonella/microsome test. *Carcinogenesis* **2** (4), 283–298.

Dobrzynska, I., Skrzydlewska, E., and Figaszewski, Z. (1999) Influence of trolox derivative and N-acetylcysteine on surface charge density of erythrocytes in methanol intoxicated rats. *Environ. Toxicol. Pharmacol.* **8**, 15–21.

European Food Safety Authority (EFSA) (2006) Opinion of the scientific panel on food additives, flavoring, processing aid and material in contact with food (AFC) on a request from the commission related to a new long term carcinogenicity study on aspartame. *EFSA J.* **356**, 1–44.

EPA (2012) Announcement: Update on Ramazzini Institute data in IRIS assessments March 9 (http://www.epa.gov/iris/ramazzini.htm).

Forbes, P.D., Urabach, F., and Davies, R.E. (1979) Enhancement of experimental photocarcinogensis by topical retinoic acid. *Cancer Lett.* **7**, 85–90.

Halliwell, B. and Gutteridge, J.M. (2007) *Free Radicals in Biology and Medicine*. Oxford University Press, Oxford.

Heidelberger, C., Freeman, A.E., Pienta, R.J., Sivak, A., Bertram, J.S., Castro, B.C., Dunkel, V.C., Francis, M.W., Kakunaga, T., Little, J.B., and Schechtman, L.M. (1983) Cell transformation by chemical agents – A review and analysis of the literature. A report of US EPA Gene Tox Program. *Mutat. Res.* **114**: 283–385.

Horton, V.L., Higochi, M.I., and Ricket, D.E. (1992) Physiologically based pharmacokinetic model for methanol in rats, monkeys and humans. *Toxicol. Appl. Pharmacol.* **117** (1), 26–36.

IPCS (1997) *Environmental Health Criteria 196 – Methanol*, World Health Organization, Geneva.

Jiang, J., Xu, Y., and Klaunig, J.E. (1998) Induction of oxidative stress in rat brain by acrylonitrile (ACN). *Toxicol. Sci.* **46** (2), 331–334.

Kadiiska, M.B. and Mason, R.P. (2000). Acute methanol intoxication generates free radicals in rats: an ESR spin trapping investigation. *Free. Radic. Biol. Med.* **28**, 1106–1114.

Kanhtala, P., Kaupplia, S., Puistola, U., and Jukkola-Vuorinen, A. (2011) Divergent behaviour of oxidative stress markers 8-hydroxydeoxyguanosine (8-OhdG) and 4 hydroxy-2-noneal (HNE) in breast carcinogensis. *Histopathology* **58** (6), 854–882.

Klaunig, J.E. and Kamendulis, L.M. (2004) The role of oxidative stress in carcinogenesis. *Annu. Rev. Pharmacol. Toxicol.* **44**, 239–267.

Klaunig, J.E., Xu, Y., Isenburg, J.S., Bachowski, S., Jiang, J., Stevenson, D.E., and Walborg, E.F. (1998) The role of oxidative stress in chemical carcinogensis. *Environ. Health Perspect.* **106** (Supp 1), 289–295.

Loft, S. and Poulsen, H.E. (1996) Cancer risk and oxidative DNA damage in man. *J. Mol. Med.* **74** (6), 297–312.

Lu, K., Gul, H., Upton, P.B., Moeller, B.C., and Swenberg, J.A. (2012) Formation of hydroxymethyl DNA adducts in rats orally exposed to stable isotope label methanol. *Toxicol. Sci.* **126** (1), 28–38.

Masten, S. e-mail Methanol Bioassay to John Clary. November 14, 2006.

McCallum, G.P., Siu, M., Ondovcik, S.L., Sweeting, J.N., and Wells, P.G. (2011a) Methanol exposure does not lead to accumulation of oxidative DNA damage in bone marrow and spleen of mice, rabbits or primates. *Mol. Carcinog.* **50**, 163–172.

McCallum, G.P., Siu, M., Sweeting, J.N., and Wells, P.G. (2011b) Methanol exposure does not produce oxidatively damaged DNA in lung, liver or kidney of adult mice, rabbits or primates. *Toxicol. Appl. Pharmacol.* **250**, 147–153.

McGregor, D.B., Edwards, I., Riach, C.G., Martin, R., Mitchell, A., and Caspary, W.J. (1988) Studies of an S9 based metabolic activation system used in the mouse lymphoma L5178Y cell mutation assay. *Mutagenesis* **3** (6), 485–490.

NEDO (1986) *Toxicological Research of Methanol as a Fuel for Power Stations*. New Energy Development Organization, Tokyo, Japan.

NEDO (1987) *Toxicological Research of Methanol As A Fuel for Power Station: Summary Report on Tests with Monkeys, Rats And Mice*. New Energy Development Organization, Tokyo, Japan.

Nelson, B.K., Brightwell, W.S., MacKenzie, D.R., Khan, A., Burg, J.R., Weigel, W.W., and Goad, P.T. (1985) Teratological assessment of methanol and ethanol at high inhalation levels in rats. *Fundam. Appl. Toxicol.* **5**, 727–736.

NTP (National Toxicology Program). (1988) Toxicology and carcinogensis studies of malonaldehye, sodium salt (3-hydroxy-2-propenal, sodium salt) in F344/N rats and B6C3F1 mice. *Technical Report* No. **331**.

NTP (2004) Pathology Working Group (PWG) chairperson's report: Lifetime study in rats conducted by Ramazzini Foundation. November 30, 2004.

NTP (2010) Letter to Bucher JR, from Marlarkey D, Herbert R, Nyska A, Sutphin Pernicka K., Subject Report of visit 4/25/2010-4/30/2010 to Ramazzini Institute, dated June 11, 2010.

NTP & EPA (2011) Summary report of the National Toxicology Program and Environmental Protection Agency-sponsored review of pathology materials from select Ramazzini Institute Rodent Cancer Bioassays, November 29 (http://1.usa.gov/zgBliG).

O'Loughlin, K., LeValley, S., Mirsalis, J., and McGregor, J. (1992) Erythrocyte micronucleus assay in normal and folate deficient swiss mice. *Environ. Mol. Mutagen.* **19** (20), 47.

Ohno, K., Tanaka-Azuma, Y., Yoneda, Y., and Yamada, T. (2005) Genotoxicity test system based on p53r2 gene expression in human cells: examination of 80 chemicals. *Mutat. Res.* **588**, 47–57.

Parthasarathy, N.J., Kumar, R.S., Manikandan, S., and Devi, R.S. (2006) Methanol-induced oxidative stress in rat lymphoid organs. *J. Occup. Health* **48**, 20–27.

Roberts, R.A., Laskin, D.L., Smith, C.V., Robertson, F.M., Allen, E.M.G., Doorn, J.A., and Slikker, W. (2009) Nitrative and oxidative stress in toxicology and disease. *Toxicol. Sci.* **112**, 4–16.

Rogers, J.M., Mole, M.L., Chernoff, N., Barbee, B.D., Turner, C.I., Logsdon, T.R., and Kavlock, R.J. (1993) The developmental toxicity of inhaled methanol in the CD-1 mouse, with quantitative dose-response modeling for estimation of benchmark doses. *Teratology* **47**, 175–188.

Schoeb, T.R., McConnell, E.E., Juliana, M.M., Davis, J.K., Davidson, M.K., and Lindsey, J.R. (2009) Mycoplasma pulmonis and lymphoma in bioassay in rats. *Vet. Pathol.* **46**, 952–959.

Shimizu, H., Suzuri, Y., Takemura, N., Goto, S., and Matsushita, H. (1995). The results of microbial mutation test for forty three industrial chemicals. *Jpn. J. Ind. Health.* **27**, 400–419.

Simmon, V.E., Kauhanen, K., and Tardiff, R.G. (1977) Mutagenic activity of chemicals found in drinking water. *Dev. Toxicol. Environ. Sci.* **2**, 249–258.

Skrzydlewska, E. and Farbiszewski, R. (1996) Diminished antioxidant defense potential of liver, erythrocytes and serum, from rats with subacute methanol intoxication. *Vet. Human Toxicol.* **38**, 429–433.

Skrzydlewska, E. and Farbiszewski, R. (1998) Lipid peroxidation and antioxidant status in the liver, erythrocytes and serum of rats after methanol intoxication. *J. Toxicol. Environ. Health* **53** (8), 637–649.

Skrzydlewska, E., Witek, A., and Farbiszewski, R. (1998) The comparison of the antioxidant defense potential of brain to liver of rats after methanol ingestion. *Comp. Biochem. Physiol.* **120**, 289–294.

Soffritti, M., Belpoggi, F., Cevolani, D., Guarino, M., Padovani, M., and Maltoni, C. (2002) Results of long-term experimental studies on the carcinogenicity of methyl alcohol and ethyl alcohol in rats. *Ann. N.Y. Acad. Sci.* **982**, 46–69.

Ward, J.M. and Alden, C.L. (2009) Confidence in rodent bioassays. *Vet. Pathol.* **46**, 790–791.

Wells, P.G., Kim, P.M., Laposa, R.R., Nicol, C.J., Parman, T., and Winn, L.M. (1997) Oxidative damage in chemical teratogenesis. *Mut. Res.* **396**, 65–78.

Wells, P.G., Bhuller, Y., Chen, C.S., Jeng, W., Kasapinovic, S., Kennedy, J.C., Kim, P.M., Laposa, R.R., McCallum, G.P., Nicol, C.J., Parman, T., Wiley, M.J., and Wong, A.W. (2005) Molecular and biochemical mechanisms in teratogenesis involving reactive oxygen species. *Toxicol. Appl. Pharmacol.* **207**, 354–366.

Wells, P.G., Lee, C.J.J., McCallum, G.P., Perstin, J., and Harper, P.A. (Eds.) (2009a) *Receptor- and Reactive Intermediate-Mediated Mechanisms of Teratogenesis*, Vol. **196**. Springer, Heidelberg, pp. 131–162, Chapter 6.

Wells, P.G., McCallum, G.P., Chen, C.S., Henderson, J.T., Lee, C.J., Perstin, J., Preston, T.J., Wiley, M.J., and Wong, A.W. (2009b) Oxidative stress in developmental origins of disease: teratogenesis, neurodevelopmental deficits, and cancer. *Toxicol. Sci.* **108**, 4–18.

INDEX

Accidental ingestion 54
Adulterated alcohol 56
  methanol instead of ethanol 57
Animal toxicity 73
Animal toxicity (Acute) 74
  dermal irritation 86
  dermal toxicity 78
  eye irritation 87
  inhalation toxicity (one exposure) 79
    kormone effects 81
  intraperitoneal toxicity 1 82
  intravenous toxicity 82
  neurobehavioral toxicity 1 77
  oral toxicity 75, 193
  RD 50 in mice 81

Carcinogenic potential and oxidative stress 176, 207, 223, 233
  hydroxynonenal-histidine protein adducts 222, 234, 266
  oxidatively damaged DNA 208, 233, 272
  rodent studies 209

Dietary exposure to methanol 48
  alcoholic beverage 48
  aspartame 48
  dimethyl dicarbonate 49
  fruits and vegetable 48

Environmental 11
  adsorption to soil materials 31
  air/water partitioning 16
partitioning in the environment 14
    soil/water partitioning 18
  availability of electron acceptors 25
  availability of nutrients for methanol breakdown 27
  biodegradation of methanol in environment 34
  biodegradation under various redox conditions 28
  commingling/cosolvency effects 21, 32
  environmental fate of methanol 11,14
  indigenous methanol-degrading microbes 24
  manufacture of industrial chemicals 49
  methanol dissolution 19
  methanol environmental release scenarios 11
    rail car or tank truck release 12
    ship or barge release.13
  methanol vs gasoline and benzene 18
  pH and temperature levels for microbial growth 27

The Toxicology of Methanol, First Edition. Edited by John J. Clary.

partitioning in the environment (*Continued*)
storage/fueling facility release 13
surface water release of methanol 33
abiotic degradation in surface water 35
bioaccumulation 35
biodegradation in surface water 34
losses from surface water 33
sources of methanol in surface water 33
volatilization 36
transport of methanol in the environment 23
losses from soil and groundwater 23
soil and/or groundwater release 23
sources in soil and groundwater 23
vaporization from soil 32
volatilization from groundwater 32
Enzymes in methanol metabolism 179
ADH activity throughout development 182
alcohol dehydrogenase (ADH) 179
catalase (CAT) 182
antioxidative role (scavenging H_2O_2) 183
peroxidative role (alcohol to aldehydes) 183
cytochrome P450 (CYP) 2E1 184
enzyme kinetic parameter for ADH 181
folate dependent dehydrogenase 187
formaldehyde dehydrogenase 186

Folate deficiency and methanol developmental toxicity 129
developmental neurotoxicity study in rats 131
dietary folic acid in mice 130
liver and plasma folate level 130
hepatic terahydrofolate in rats 133
Folate deficient rats 92, 109

History of methanol production 1
Hormones levels in male rats 108
Human (controlled exposure) 60
blood and urine methanol levels 61
chronic inhalation workplace exposure 63
elimination half-life-62
neurobehavioral effects 60
urinary and serum formate 61
Human (dermal exposure) 59
absorption through skin 59
skin effects 59
Human (environmental exposure) 49 11, 12, 13
Human Cancer Data 209, 276
Human exposure (occupational exposure) 1 50, 54

In utero methanol exposure in humans 62

Management, methanol poisoning 64
American Academy of Clinical Toxicology Practice guidelines 64
fomepizole 64, 193
hemodialysis 64
sodium bicarbonate 64
Methanol additives 37
bitrex (bitterant) 38
luminosity 38

Methanol, commercial uses
chemical building block 5
energy/fuel 6
California methanol fuel program 7
China M-15, M-30, M85, M-100 fuel use 8
Methanol *vs.* ethanol 172, 228
embryo culture 228
free radical spin trapping 232
genetic modulation of catalase 228
Methanol induced oxidative stress 82, 171, 273
acute rodent m studies 82, 273
effects of antioxidants 85, 274
free radical and electron spin resonance spectroscopy 86, 274
lipid peroxidation 83, 221, 227, 273
malonaldehyde (MDA) 84, 221, 227
oxidative stress 82
radical oxygen species (ROS) 83, 177, 273
species differences 77, 194
thiobarbituric acid reactive substance (TBA) 83, 221
tissue effected (liver, erythrocyte, serum, brain, lymphoid organs) 85
Metabolic pathways in mice and humans 155, 171
Metabolism of methanol in humans 50
Methanol, physical and chemical properties 18

Neurodevelopmental effects of methanol 204

Normal metabolism in humans 50
alcohol dehydrogenase (ADH) 50
formaldehyde 50
formate - carbon dioxide and water 50
formic acid 50
low level in blood salvia and expired breath 50

Odor threshold 50
Over exposure to methanol in humans 51
built up of formate and hydrogen ion 51
saturation of formic acid dehydrogenase 51
Oxidative stress mechanisms 213
DNA oxidation 219
embryonic drug exposure and reactive oxygen species (ROS) formation 213
lipid peroxidation 220
macromolecular damage 217
protein oxidation 218
signal transduction 214

Pathogenesis, birth defects 127
comparison of mouse strains 129
in vitro studies 128
in vivo studies 127
mouse embryo 128
mouse inhalation 127
mouse oral 127
PBPK models in mice and humans 145
background 145
Battelle's Pacific Northwest National laboratory PBPK model 147
blood methanol levels in pregnant mice 146

PBPK models in mice and humans (*Continued*)
Boucher's PBPK model 151
metabolic pathways in mice and humans 155
mouse and human blood methanol concentration 151
PBPK model parameters 150
visual optimization weakness 161
metabolic difference in mice and humans 146
Pharmacokinetics, methanol and formic acid 188
mice stains (CD1, C57BL/6J and C#H) 189
mice, rabbits and primate 179. 188
routes of exposure 190
Possible cancer mechanisms 270
genotoxicity 270
in vitro 270
in vivo 270
oxidative damage 176, 177, 208, 223, 272
rodent special studies 273
short term studies in mice, rabbits, primates 274
Production methods, current 1
other production sources 3
biomass production 4
coal gasification 3
primary method steam reformation of natural gas 2
Production methods, new 8
captured atmospheric CO_2 8
renewable generated hydrogen 8

Rodent cancer bioassay 256
dermal cancer bioassay 268
mice 268 thirty-week study - 268
cancer incidence 268
life time (3 times/week) (eppley) 269
cancer incidence 270
inhalation cancer bioassays 266
mice 268 eighteen month (NEDO study) 268
cancer incidence 268
rats 266 Two years (NEDO study) 266
cancer incidence 267
oral cancer bioassays
mice 263 life time MDA (Eppley study) 263
cancer incidence 265
rats 256 (lifetime study drinking water (Ramazzini Institute) 256
cancer incidence 257
possible problems
histopathological diagnosis 259
infection 261
Role of methanol and metabolites in the developmental toxicity 133
embryo culture studies 133, 135
formate levels following inhalation in mice 135
formaldehyde possible role 136

Sensitization studies 87
Short term special studies in mice, rabbits, primates 273
evidence for MeOH-initiated ROS formation 223
mechanism of MeOH-initiated ROS formation 227
Species differences in Methanol metabolism 171, 178, 179
dose and route of exposure 179

Strain differences in methanol toxicity 189, 194
  acute metabolic acidosis, ocular toxicity and death 191
Subcutaneous studies 82
Symptoms over exposure to methanol 51, 192
  level of methanol in blood and urine 52, 193
  methanol levels in breath 53
  mortality 52, 192
  pathological changes 57
  urinary formic acid 53
  visual disturbances 51, 192

Toxicity, aquatic
  acute fish toxicity 96
  acute toxicity to invertebrates 98
  chronic toxicity 98
Toxicity, chronic dermal 96
  mice 96
Toxicity, developmental 110, 172
  mice 115
    early pregnancy studies 117
    intraperitoneal 119
    oral 117
    teratology inhalation studies 115, 116
  mice stains (CD1, C57BL/6J and C3H) 195
  mouse embryo 198, 228
  non-human primates 120, 194
    inhalation developmental and reproductive 120, 194
    inhalation neurobehavioral infants 123
  rabbits 194
  rats 111
    early pregnancy studies 115
    2 generation studies 112
    postnatal 112
    prenatal 112
    teratology inhalation studies 111
Toxicity, inhalation (chronic)
  mice 93
  rats 91
Toxicity, inhalation (non human primates) 87
  pilot studies 87
  subacute 90
  subchronic 88
Toxicity, inhalation (subacute) 87
  rats 90, 92
Toxicity, inhalation (subchronic)
  dogs 94
  folate deficient rats 92, 109
  rats 92
Toxicity, oral - 94
  mice 95
  non human primates 95
  rats 94
Toxicology, reproductive (animals) 110
Toxicology, reproductive (humans) 108

www.ingramcontent.com/pod-product-compliance
Lightning Source LLC
Chambersburg PA
CBHW071301180926
51613CB00001B/100
9780470317594